Convergence of IoT, Blockchain, and Computational Intelligence in Smart Cities

This edited book presents an insight for modelling, procuring, and building the smart city plan using the Internet of Things (IoT) and a security framework using blockchain technology. The applications of Li-Fi and 5G in smart cities are included, along with their implementation, challenges, and advantages. This book focuses on the use of IoT and blockchain in the day-to-day transparent and recorded activities of citizens of smart cities like, smart citizen management. The future for upgrading the system as per technological advancements is also discussed. This book:

- integrates IoT, blockchain, Li-Fi, and 5G in smart city implementation
- covers smart supply chain management using IoT
- outlines the state-of-the-art and sustainable implementation of smart cities and practical challenges
- includes sustainable development of smart cities
- presents detailed explanation of case studies of smart cities of developed countries and developing countries and their comparisons

This book is aimed at researchers and graduate students in Artificial Intelligence, Urban Planning, and Information Technology Systems and Management.

Dr. Rajendra Kumar is presently working as an Associate Professor of Computer Science at Sharda University, Greater Noida, Uttar Pradesh. He is vice president and a life member of the Society for Research Development (SRD) and member of IEEE, CSI, IACSIT, IAENG, UACEE, SCIEI, CSTA, etc.

Dr. Vishal Jain is presently working as an Associate Professor at Sharda University, Greater Noida, Uttar Pradesh. He received a Young Active Member Award for the year 2012–13 from the Computer Society of India, Best Faculty Award for the year 2017, and Best Researcher Award for the year 2019 from BVICAM, New Delhi.

Dr. Leong Wai Yie is currently working as Pro Vice Chancellor at INTI International University, Kuala Lumpur, Malaysia. Wai Yie is currently the Immediate Past

Chairperson of the Institution of Engineering and Technology (Malaysia Local Network) and the Vice President of the Institution of Engineers Malaysia (IEM).

Dr. Sunantha Teyarachakul is an Associate Professor at California State University, Fresno, CA, United States. She is also a member of the Institute for Operations Research and the Management Sciences (INFORMS).

Computational Intelligence Techniques
Series Editor: Vishal Jain

The objective of this series is to provide researchers a platform to present state of the art innovations, research, and design and implement methodological and algorithmic solutions to data processing problems, designing and analyzing evolving trends in health informatics and computer-aided diagnosis. This series provides support and aid to researchers involved in designing decision support systems that will permit societal acceptance of ambient intelligence. The overall goal of this series is to present the latest snapshot of ongoing research as well as to shed further light on future directions in this space. The series presents novel technical studies as well as position and vision papers comprising hypothetical/speculative scenarios. The book series seeks to compile all aspects of computational intelligence techniques from fundamental principles to current advanced concepts. For this series, we invite researchers, academicians and professionals to contribute, expressing their ideas and research in the application of intelligent techniques to the field of engineering in handbook, reference, or monograph volumes.

Smart Computing and Self-Adaptive Systems
Simar Preet Singh, Arun Solanki, Anju Sharma, Zdzislaw Polkowski and Rajesh Kumar

Advancing Computational Intelligence Techniques for Security Systems Design
Uzzal Sharma, Parmanand Astya, Anupam Baliyan, Salah-ddine Krit, Vishal Jain and Mohammad Zubair Kha

Graph Learning and Network Science for Natural Language Processing
Edited by Muskan Garg, Amit Kumar Gupta and Rajesh Prasad

Computational Intelligence in Medical Decision Making and Diagnosis: Techniques and Applications
Edited by Sitendra Tamrakar, Shruti Bhargava Choubey and Abhishek Choubey

Applications of 5G and Beyond in Smart Cities
Edited by Ambar Bajpai and Arun Balodi

Healthcare Industry 4.0: Computer Vision-Aided Data Analytics
Edited by P. Karthikeyan, Polinpapilinho F. Katina and R. Rajaagopal

Blockchain Technology for IoE: Security and Privacy Perspectives
Edited by Arun Solanki, Anuj Kumar Singh, and Sudeep Tanwar

Convergence of IoT, Blockchain and Computational Intelligence in Smart Cities
Edited by Rajendra Kumar, Vishal Jain, Leong Wai Yie, and Sunantha Teyarachakul

Intelligent Techniques for Cyber-Physical Systems
Edited by Mohammad Sajid, Anil Kumar, Jagendra Singh, Osamah Ibrahim Khalaf, and Mukesh Prasad

For more information about this series, please visit: www.routledge.com/Computational-Intelligence-Techniques/book-series/CIT

Convergence of IoT, Blockchain, and Computational Intelligence in Smart Cities

Edited by
Rajendra Kumar, Vishal Jain, Leong Wai Yie, and
Sunantha Teyarachakul

CRC Press
Taylor & Francis Group
Boca Raton London New York

CRC Press is an imprint of the
Taylor & Francis Group, an **informa** business

Designed cover image: Shutterstock

First edition published 2024
by CRC Press
2385 NW Executive Center Drive, Suite 320, Boca Raton FL 33431

and by CRC Press
4 Park Square, Milton Park, Abingdon, Oxon, OX14 4RN

CRC Press is an imprint of Taylor & Francis Group, LLC

© 2024 selection and editorial matter, Rajendra Kumar, Vishal Jain, Leong Wai Yie and Sunantha Teyarachakul; individual chapters, the contributors

Reasonable efforts have been made to publish reliable data and information, but the author and publisher cannot assume responsibility for the validity of all materials or the consequences of their use. The authors and publishers have attempted to trace the copyright holders of all material reproduced in this publication and apologize to copyright holders if permission to publish in this form has not been obtained. If any copyright material has not been acknowledged please write and let us know so we may rectify in any future reprint.

Except as permitted under U.S. Copyright Law, no part of this book may be reprinted, reproduced, transmitted, or utilized in any form by any electronic, mechanical, or other means, now known or hereafter invented, including photocopying, microfilming, and recording, or in any information storage or retrieval system, without written permission from the publishers.

For permission to photocopy or use material electronically from this work, access www.copyright.com or contact the Copyright Clearance Center, Inc. (CCC), 222 Rosewood Drive, Danvers, MA 01923, 978-750-8400. For works that are not available on CCC please contact mpkbookspermissions@tandf.co.uk

Trademark notice: Product or corporate names may be trademarks or registered trademarks and are used only for identification and explanation without intent to infringe.

ISBN: 978-1-032-40424-0 (hbk)
ISBN: 978-1-032-40426-4 (pbk)
ISBN: 978-1-003-35303-4 (ebk)

DOI: 10.1201/9781003353034

Typeset in Times
by KnowledgeWorks Global Ltd.

Contents

List of Figures and Tables xv
Contributors xxi
Preface xxv
Acknowledgements xxvii

Chapter 1   5G Intelligent Transportation Systems for Smart Cities 1

Wai Yie Leong and Rajendra Kumar

     1.1   Introduction 1
     1.2   Intelligent Transportation System 1
           1.2.1   Types of Intelligent Transportation Systems 2
     1.3   Data Analytics on Vehicle Connections 2
     1.4   Evolution and History 5
     1.5   How Linked Vehicles Function 7
     1.6   Smart Car 8
     1.7   Fundamental Autonomous Vehicle Technology 8
     1.8   The Intelligent Driver Model and Approach 9
     1.9   Vehicles Connectivity and Networking 10
     1.10   Enabling and Facilitating Technologies 11
           1.10.1   Environmental State 11
           1.10.2   Tracking Lane 12
           1.10.3   Identifying and Detecting Traffic Signals 13
           1.10.4   Cars Detection 14
           1.10.5   Adaptive Cruise Monitor and Control 14
     1.11   Modulation System for Intelligent Cars 15
     1.12   Spatial Modulation (SM) Advantages and Downsides 15
     1.13   The Development of Industry Revolution 5.0 16
     1.14   Significant Technologies in Intelligent
               Transportation Systems 17
     1.15   Smart Transportation Structures and Architectures 18
     1.16   Applications for Intelligent Transportation 20
     1.17   Conclusions 23
     References 23

Chapter 2   IoT Smart Environment: A Comprehensive Study on
Challenges and Solutions 26

*M. R. Sundarakumar, S. Sankar, Somula Ramasubbareddy,
Velleangiri Vinodhini, and Balusamy Balamurugan*

     2.1   Introduction 26
     2.2   IoT Challenges 26

	2.3	IoT Characteristics	27
	2.4	IoT Process Management	29
	2.5	IoT Challenges and its Solutions	29
		2.5.1 Analytics of Edge Computing	30
		2.5.2 IoT Security Issues	31
		2.5.3 Industrial IoT Challenges	31
		2.5.4 IoT Cloud Computing Platforms	32
		2.5.5 ML/AI-Based IoT Applications	33
		2.5.6 IoT in Sustainability and Climate Change	34
		2.5.7 IoT with Blockchain	34
		2.5.8 Services of Cloud and Edge Computing in IoT	35
		2.5.9 IoT Devices – Sensors Increasingly Becoming Commodity	36
		2.5.10 LPWAN and 5G Technologies for IoT	37
	2.6	IoT Future Trends	37
	2.7	Conclusion	37
	References		38

Chapter 3 Enabling Technologies for Internet of Things (IoT)-Based Smart Cities 41

Velleangiri Vinodhini, B. Sathiyabhama, S. Sankar, S. Ramasubbareddy, M. R. Sundarakumar, and B. Balamurugan

	3.1	Introduction	41
	3.2	Overview of IoT Architecture	42
	3.3	Enabling Technologies	43
		3.3.1 Things Layer	43
		3.3.2 Connectivity Layer	44
		3.3.3 The Global Infrastructure Layer	46
		3.3.4 Data Ingestion Layer	51
		3.3.5 Data Analysis	52
		3.3.6 Application Layer	52
		3.3.7 People and Process Layer	53
	3.4	Conclusion	55
	References		56

Chapter 4 Reshape the Sustainable State-of-the-Art Development of Smart Cities 58

Tarana Afrin Chandel

	4.1	Introduction	58
		4.1.1 Concept of Smart City Model	59
		4.1.2 Sustainability a Strategic Plan of Smart Cities	59
		4.1.3 Technology-Based Smart City	60
		4.1.4 Sustainable Transport for Smart Cities	63

		4.1.5	Sustainable Smart Learning	64
		4.1.6	Smart Fitness	68
		4.1.7	Smart Ecosystem	70
		4.1.8	Smart Live-Style	72
		4.1.9	Smart Finance	73
		4.1.10	Smart Governance	75
	4.2	Conclusion		76
	References			77

Chapter 5 Emergence of Big Data and Blockchain Technology in Smart City 83

Meenu Gupta, Rakesh Kumar, Anamika Larhgotra, and Chetanya Ved

	5.1	Introduction	83
	5.2	Literature Survey	84
	5.3	Smart City	85
	5.4	Emerging Features of Smart Cities	88
		5.4.1 Big Data	88
		5.4.2 Artificial Intelligence	89
		5.4.3 FinTech	90
		5.4.4 e-Governance	90
		5.4.5 5G/6G-Based Communication Backbone	91
		5.4.6 Cloud Computing	91
		5.4.7 e-Healthcare	92
	5.5	Security Challenges in Smart City	92
	5.6	Safety Challenges Tackled by Blockchain over Big Data	94
	5.7	Analysis of Blockchain Implementation with Big Data	97
	5.8	Way Forward	98
	5.9	Conclusion	99
	References		99

Chapter 6 Comprehensive Review: Recent Advancements and Applications of Cyber-Physical Systems for IoT Devices 102

Umesh Kumar Lilhore, Sarita Simaiya, Martin Margala, Prasun Chakrabarti, and Atul Garg

	6.1	Introduction	102
	6.2	Related Work	104
	6.3	Cyber-Physical System and IoT	105
		6.3.1 Physical Components of CPS	106
		6.3.2 Architecture of Cyber-Physical System	106
		6.3.3 Impacts Regarding CPS Security in IoT Devices	108
	6.4	Cybersecurity Issues and Research Gaps	108

6.5	Cyber-Physical Security Threats	109
	6.5.1 Classification of CPS Security Threats	109
6.6	Cyber-Physical System Security Solutions	111
6.7	Conclusion	112
References		112

Chapter 7 Deep Learning-Based Autonomous Driving and Cloud Traffic Management System for Smart City 117

Soujanya Syamal, Joyatee Datta, Srijita Basu, and Shuvendu Das

7.1	Introduction	117
7.2	Related Work	117
7.3	System Architecture: Autonomous Driving	118
	7.3.1 Components	118
	7.3.2 Sense	121
	7.3.3 Learn	123
	7.3.4 Act	125
7.4	System Architecture: Cloud Traffic Management	126
	7.4.1 Cloud-Based Vehicle and Traffic Control System	127
	7.4.2 Local Traffic Management System	127
	7.4.3 Global Traffic Monitoring System	128
7.5	Workflow	129
7.6	Result	131
7.7	Conclusion	133
References		133

Chapter 8 Security and Privacy Challenges in IoT System Resolving Using Blockchain Technology 136

Gauri Shankar, Gaganpreet Kaur, and Sukhpreet Kaur Gill

8.1	Introduction	136
	8.1.1 Internet of Things (IoT)	136
	8.1.2 Features of IoT	138
8.2	Applications of IoT	138
	8.2.1 IoT and Healthcare	139
	8.2.2 IoT in Industry	139
	8.2.3 Education	140
	8.2.4 e-Governance	141
8.3	Cyber-Attacks on IoT Infrastructure	141
	8.3.1 Attacks on IoT Software	142
	8.3.2 Attacks on IoT Hardware	142
	8.3.3 Types of Attacks on IoT Infrastructure	144
8.4	Solutions with Blockchain Technology	147
	8.4.1 Blockchain Technology	147
	8.4.2 Basic Architecture of Blockchain	153

		8.4.3	Blockchain in IoT Infrastructure 154

      8.4.3   Blockchain in IoT Infrastructure 154
      8.4.4   The Traditional Architecture of Smart City 154
  8.5 Conclusion 158
  References 158

Chapter 9 Secure Blockchain-Based E-voting System for
Smart Governance 162

*Raja Muthulagu, Pranav M. Pawar, Ashish Kumar Jha,
Karan Sharma, and Kavya Parthasarathy*

  9.1 Introduction 162
  9.2 Blockchain for E-voting 164
      9.2.1   Blockchain Structure 164
      9.2.2   Advantages of Using Blockchain in
              Electronic Voting 164
      9.2.3   Components of the E-voting System 166
  9.3 Related Work 167
  9.4 Proposed E-voting System Based on Smart Contracts 173
      9.4.1   Concept behind Voting System Using
              Smart Contracts 174
      9.4.2   Software and Tool Requirements 174
  9.5 Proposed E-voting System Based on
      Consensus Algorithms 175
  9.6 Implementation and Results 177
      9.6.1   Vote Casting 177
      9.6.2   After Vote Casting 178
  9.7 Conclusions 181
  References 181

Chapter 10 Design of Intelligent Healthcare Information System Using
Data Analytics 184

P. Nagaraj, V. Muneeswaran, and Pandiaraj Annamalai

  10.1 Introduction 184
  10.2 Work of Descriptive Analytics in Healthcare Systems 185
  10.3 Work of Diagnostics Analytics in Healthcare System 185
       10.3.1 Work of Drill-Down 186
       10.3.2 Work of Data Discovery 187
       10.3.3 Work of Data Mining 188
  10.4 Work of Predictive Analytics in the Healthcare System 188
  10.5 Work of Prescriptive Analytics in the Healthcare System 190
  10.6 Data Collection in the Healthcare System 191
  10.7 Data Extraction in the Healthcare System 192
  10.8 Data Generation in the Healthcare System 193
  10.9 Analysis of the Healthcare System 194

10.10 Visualization and Reporting in the Healthcare System 194
10.11 Problems, Challenges, Barriers, and Issues in the
       Healthcare System 195
10.12 Managerial Issues in the Healthcare System 197
10.13 Data Quality in the Healthcare System 197
10.14 Public Reporting Data in the Healthcare System 198
10.15 Data Privacy and Governance in the Healthcare System 199
10.16 Conclusion 200
References 201

Chapter 11 Automatic Room Light Controller Using Arduino
and PIR Sensor 203

Huma Khan, Harsh Dubey, and Yasir Usmani

11.1 Introduction 203
11.2 Literature Survey 205
11.3 IoT Components 210
11.4 Architecture of Room Automation with IoT 210
11.5 Proposed Circuit Descriptions 211
11.6 Pir Sensor and Relay Module 212
       11.6.1 Working of the System 212
       11.6.2 The Circuit Diagram 213
11.7 Results 214
11.8 Conclusion 214
11.9 Future Scope 214
References 215

Chapter 12 Role of IoT in Supply-Chain Management Processes 218

Usha Yadav and Sheetal Soni

12.1 Introduction 218
12.2 Increasing Complexities of Supply-Chain Processes 219
12.3 Evolution in Supply-Chain Environment 221
12.4 Internet of Things 222
12.5 Role of IoT in Sustainable Supply-Chain Management 222
12.6 Detailed Literature on Supply-Chain Process and IoT 223
       12.6.1 Source 223
       12.6.2 Make 224
       12.6.3 Deliver 224
       12.6.4 Return 225
12.7 RFID for Innovative Supply-Chain Management 225
       12.7.1 RFID in Integration 227
       12.7.2 RFID in Operations 228
       12.7.3 RFID in Purchasing 229
       12.7.4 RFID in Distribution 229

12.8 IoT-Based Fuzzy Logic Decision System for Reversed Logistics ... 230
    12.8.1 Input 1: Frequency of Maintenance/Repair ... 231
    12.8.2 Input 2: Time Utilization of Product ... 232
    12.8.3 Input 3: Product Inactive Mode ... 232
    12.8.4 Output: Dispose Intension ... 233
12.9 Blockchain and IoT for SCM ... 234
12.10 Conclusion ... 235
References ... 235

Chapter 13 Moving Toward Autonomous Vehicles (Drones and Robots) for Efficient and Smart Delivery of Services Using Hybrid Ontological-Based Approach ... 239

Nidhi, Jitender Kumar, and Sofia Sandhu

13.1 Introduction ... 239
13.2 Related Works on Drone and Robotics-Based Logistics ... 240
13.3 Methodology ... 243
    13.3.1 Proposed Delivery Process Using AGVs (Robots) ... 243
    13.3.2 Proposed Delivery Process Using Ontological-Based Hybrid Approach (Drones+Robots) ... 243
    13.3.3 Ontological Factor to Filter Delivery Orders in Hybrid System ... 246
13.4 Case Study and Analysis ... 246
13.5 Conclusion and Future Scope ... 248
References ... 249

Chapter 14 Relevance and Predictability in Wireless Multimedia Sensor Network in Smart Cities ... 251

Raj Gaurang Tiwari, Pratibha, Sandip Vijay, Sandeep Dubey, Ambuj Kumar Agarwal, and Megha Sharma

14.1 Introduction ... 251
14.2 Related Works ... 252
14.3 Transport Protocol ... 254
    14.3.1 Reliability Mechanism ... 254
    14.3.2 Performance Metrics ... 254
14.4 Comparison of Protocols ... 258
14.5 Conclusion ... 259
14.6 Future Scope ... 260
References ... 260

Index ... 263

List of Figures and Tables

Figure 1.1	The expectation and estimation level of automotive production between 2023 and 2030.	3
Figure 1.2	Statistics of autonomous vehicles recorded in California for testing on public roads.	3
Figure 1.3	Fatal accidents undermine trust in self-driving vehicles.	4
Figure 1.4	A statistic on US consumers buy self-driving cars.	4
Figure 1.5	Divergent views on self-driving cars around the world.	5
Figure 1.6	CACC layout.	11
Figure 1.7	Environmental state sensing.	12
Figure 1.8	Lane detection.	13
Figure 1.9	Identification of street signs for level warning broadcast.	14
Figure 1.10	Adaptive cruise control (ACC).	15
Figure 1.11	Multi-station shared vehicle system and framework.	19
Figure 1.12	Flow of multilane traffic sequence.	21
Figure 1.13	Information about emergency vehicles' approach.	22
Figure 1.14	Allocation of ASE camera displayed at Dubai Smart City.	22
Figure 2.1	IoT process.	27
Figure 2.2	IoT challenges.	28
Figure 2.3	IoT characteristics.	29
Figure 2.4	IoT process management.	30
Figure 2.5	IoT edge computing analytics.	31
Figure 2.6	IoT security issues.	32
Figure 2.7	Industry IoT challenges.	33
Figure 2.8	SaaS challenges faced in IoT.	33
Figure 2.9	IoT with ML AI techniques.	34
Figure 2.10	IoT applications on sustainability and climate change.	35
Figure 2.11	IoT applications with block chain.	36
Figure 2.12	IoT with cloud and edge computing applications.	36
Figure 3.1	The seven-layered architecture of IoT.	42

Figure 3.2	CoAP message format.	48
Figure 3.3	CoAP architecture.	48
Figure 3.4	MQTT broker architecture.	49
Figure 3.5	XMPP client and server.	50
Figure 3.6	AMQP broker architecture.	50
Figure 3.7	DDS architecture.	51
Figure 3.8	IoT security.	53
Figure 3.9	Data security.	54
Figure 4.1	World wise internet user since 2005.	60
Figure 4.2	Smart City and AI increasing from 2014 till 2021	61
Figure 4.3	Sustainable development goals for smart City	61
Figure 4.4	Green smart teaching-learning environment	65
Figure 4.5	Process of machine learning	67
Figure 4.6	Types of machine learning	67
Figure 4.7	Data-driven smart ecosystem	71
Figure 4.8	Joy-driven smart city	72
Figure 4.9	Change in global population in rural and urban from 1950 to 2050 (Source: United Nations)	73
Figure 4.10	Transformation toward urbanization region wise from 1950 to 2050 (Source: United Nations)	74
Figure 5.1	Big data management.	89
Figure 5.2	Integration of big data management with the blockchain technology.	95
Figure 5.3	DLT system.	96
Figure 6.1	Overview of CPS.	103
Figure 6.2	Physical components of CSP.	107
Figure 6.3	IoT architecture and CPS.	107
Figure 6.4	Cybersecurity schemes.	111
Figure 7.1	The SLA model.	119
Figure 7.2	Cloud architecture.	127
Figure 7.3	Cloud Traffic connection simulation.	128
Figure 7.4	Workflow.	129

Figure 7.5	CNN deep learning layers.	130
Figure 7.6	CNN model.	131
Figure 7.7	Simulation graph.	132
Figure 8.1	Essential technologies of IoT.	137
Figure 8.2	Applications of IoT.	139
Figure 8.3	Types of hardware attacker tools.	143
Figure 8.4	Types of attacks in recent years and number victims.	145
Figure 8.5	Types of attacks in recent years and losses in millions.	146
Figure 8.6	Key elements of blockchain.	148
Figure 8.7	Public-key cryptography.	151
Figure 8.8	Merkle tree.	152
Figure 8.9	Blockchain architecture.	153
Figure 8.10	Layers in IoT infrastructure.	155
Figure 8.11	IoT infrastructure of smart city.	155
Figure 8.12	Entities of cloud storage in IoT infrastructure of smart city.	156
Figure 8.13	Blockchain-based IoT Infrastructure of smart city.	157
Figure 8.14	Communication between blockchain-based IoT infrastructures of smart city.	158
Figure 9.1	Blockchain creation wherein a new transaction is added to the blockchain.	165
Figure 9.2	Blockchain network architectures.	166
Figure 9.3	Blockchain and smart contract.	173
Figure 9.4	Proposed voting system using blockchain and smart contracts.	174
Figure 9.5	Working of voting system using smart contracts.	177
Figure 9.6	The user has cast the vote.	179
Figure 9.7	Voter list.	180
Figure 10.1	Different types of data roles in the healthcare system.	185
Figure 10.2	Examining the patient "what happened?" to him/her.	186
Figure 10.3	Examining the patient "why it happened?".	187
Figure 10.4	Techniques of predictive analytics.	189
Figure 10.5	Checking the patient data by taking the predictive analysis.	190

Figure 10.6	Prescriptive analysis techniques based on the healthcare system.	191
Figure 10.7	Data collection process is done in the healthcare system	192
Figure 10.8	Data extraction is done in the healthcare system.	193
Figure 10.9	Data generation is done by the healthcare system.	194
Figure 10.10	Analysis of data done by the healthcare system.	195
Figure 10.11	Problems and challenges faced in the healthcare system.	196
Figure 10.12	Management issues faced by the healthcare system.	197
Figure 10.13	Data quality is done in the healthcare system.	198
Figure 10.14	Examples of public reporting in the healthcare system.	199
Figure 10.15	Data governance and security did in the healthcare system	200
Figure 11.1	Arduino UNO.	204
Figure 11.2	Arduino UNO board pin index.	209
Figure 11.3	Architecture of room automation with IoT.	211
Figure 11.4	PIR sensor.	212
Figure 11.5	Arduino working with PIR.	213
Figure 12.1	Challenges in supply-chain management.	219
Figure 12.2	Five major challenges faced by supply-chain leaders.	220
Figure 12.3	Usage of RFID in supply-chain management.	228
Figure 12.4	Conceptual framework of fuzzy inference system.	230
Figure 12.5	Fuzzy Set of $\mu_F(f_i)$.	231
Figure 12.6	Fuzzy set of $\mu_U(u_i)$.	232
Figure 12.7	Fuzzy Set of $\mu_M(m_i)$.	233
Figure 12.8	Fuzzy Set of $\mu_{DI}(di_i)$.	233
Figure 12.9	Basic structure of a blockchain.	234
Figure 13.1	Food delivery process using robots	244
Figure 13.2	Food delivery process using hybrid approach	245
Figure 13.3	Delivery performance using robots	247
Figure 13.4	Delivery performance using drones and robots.	248
Figure 14.1	Conventional wireless sensors network.	252
Figure 14.2	Multimedia data transmission in accordance with the type of data.	256

Figure 14.3	Flowchart of multimedia data transmission.	257
Table 3.1	IoT Protocol Standards	44
Table 3.2	IoT Data Exchange Protocol and IoT Platforms	47
Table 3.3	Security Process	54
Table 3.4	Issues in Data Security	55
Table 4.1	Involvement of AI in Various Domain of SDG for Smart City	62
Table 5.1	Associated Research Papers	86
Table 6.1	Comparison of Review on CPS Risk, Challenges, and Security of IoT Devices	105
Table 6.2	Classification of CPS Security Threats	109
Table 8.1	Types of Software Attacks on IoT with Exploitation Tool	142
Table 8.2	Types of Attacks on IoT Architectures	146
Table 8.3	Countermeasures against Attacks on the IoT Infrastructure, Using Blockchain Technology	156
Table 12.1	Literature Review of Supply-Chain Process and IoT	226
Table 13.1	An Overview of Relevant Works on Drone and Robotics Logistics Systems	242
Table 14.1	Existing Transport Protocols Reliability Mechanism and Energy Efficiency Comparison	259
Table 14.2	Existing Transport Protocols Congestion Control Comparison	259

Contributors

Nidhi
DCRUST
Murthal, India

Pratibha
Shri Ramswaroop Memorial College
  of Engineering and Management
Lucknow, India

Ambuj Kumar Agarwal
School of Engineering & Technology
Sharda University
Greater Noida, India

Balamurugan Balusamy
Shiv Nadar University
Greater Noida, India

Srijita Basu
Jadavpur University
Kolkata, India

Prasun Chakrabarti
Sir Padampat Singhania University
Udaipur, India

Tarana Afrin Chandel
Integral University
Lucknow, India

Shuvendu Das
Chandigarh University
Mohali, India

Joyatee Datta
Institute of Engineering and
  Management
Kolkata, India

Harsh Dubey
Galgotias University
Greater Noida, India

Sandeep Dubey
Shri Ramswaroop Memorial
  College of Engineering and
  Management
Lucknow, India

Atul Garg
Chitkara University
Rajpura, India

Sukhpreet Kaur Gill
Canadore College
Ontario, Canada

Meenu Gupta
Chandigarh University
Mohali, India

Ashish Kuma Jha
Carnegie Mellon University
Silicon Valley, Northern
  California

Gaganpreet Kaur
Chitkara University
Institute of Engineering
  and Technology
Chitkara University
Rajpura, India

Huma Khan
Galgotias University
Greater Noida, India

Rakesh Kumar
Chandigarh University
Mohali, India

Jitender Kumar
DCRUST
Murthal, India

Rajendra Kumar
School of Engineering & Technology,
  Sharda University
Greater Noida, India

Anamika Larhgotra
Chandigarh University
Mohali, India

Wai Yie Leong
INTI International University
Kuala Lumpur, Malaysia

Umesh Kumar Lilhore
Chandigarh University
Mohali, India

Martin Margala
University of Louisiana
Lafayette, Louisiana

V. Muneeswaran
Kalasalingam Academy of
  Research and Education
Srivilliputhur, India

Raja Muthalagu
Birla Institute of Science and
  Technology Pilani
Dubai Campus, Dubai

P. Nagaraj
Kalasalingam Academy of Research
  and Education
Srivilliputhur, India

Annamalai Pandiaraj
SRM Institute of Science and
  Technology
Chennai, India

Kavya Parthasarathy
Birla Institute of Science and
  Technology Pilani
Dubai Campus, Dubai

Pranav M. Pawar
Birla Institute of Science and
  Technology Pilani
Dubai Campus, Dubai

Sundara Kumar M. R.
Galgotias University
Greater Noida, India

Somula Ramasubbareddy
VNR Vignana Jyothi
Institute of Engineering &
  Technology
Hyderabad, India

S. Sankar
Sona College of Technology
Salem, India

Sofia Sandhu
University of Scotland
Glasgow, Scotland

B. Sathiyabhama
Sona College of Technology
Salem, India

Gauri Shankar
Chandigarh University
Ludhiana, India

Karan Sharma
Birla Institute of Science and
  Technology Pilani
Dubai Campus, Dubai

Megha Sharma
GNIOT-IPS
Greater Noida, India

Sarita Simaiya
Chandigarh University
Mohali, India

Sheetal Soni
National Institute of Fashion Technology
Jodhpur, India

Soujanya Syamal
Cranfield University
Bedfordshire, Oxfordshire

Raj Gaurang Tiwari
Chitkara University Institute of Engineering and Technology
Chitkara University
Rajpura, India

Chetanya Ved
Bharati Vidyapeeth's College of Engineering
New Delhi, India

Sandip Vijay
Shivalik College of Engineering
Dehradun, India

Velleangiri Vinodhini
Sona College of Technology
Salem, India

Usha Yadav
National Institute of Fashion Technology
Bengaluru, India

Yasir Usmani
Galgotias University
Greater Noida, India

Preface

Nowadays, the majority of individuals use cellphones and electronic payments for a variety of daily activities and transactions. The concept of smart city will enable the users to operate their devices, track various services, and receive important alerts on their mobile phones, and they will be able to take required action on time. With the implementation of smart city, majority of controls of required services will be on users' smartphones. It will save time and money both. This book will provide readers a framework for managing resource conservation in smart cities and the optimal use of resources and services through time and effort savings manner. Also, it aims to provide improved physical and financial security, and the early detection of unpleasant acts using sensors in order to prevent them. The applicability of a smart city applies to everyone, including farmers who practise smart farming, front-line employees who serve in a healthy manner, those who pay taxes to support the government economy, and those who receive benefits from government programmes in a transparent manner.

Our lives are becoming more comfortable and tangible as a result of smart technologies. In light of that, cities are becoming "smart cities" by implementing the Internet of Things (IoT) effectively. Utilizing IoT makes it possible to use wireless sensor networks; however, each node of the IoT system is vulnerable to various threats. The Internet of Everything (IoE) and the Internet of Computing (IoC) are combined to enable greater connectivity across a range of communication channels. Blockchain suggestively uses mathematical hashing and encryption tools for data security. By achieving innovative solutions, the citizens of the smart city will be able to collaborate with local governance for various services but not limited to subsidies under government schemes, health and safety, smart assets, education and employment, real estate, waste and sanitation, and renewable energy.

To fulfil the said requirement, it is highly desirable to integrate and manage physical and societal infrastructures, ensuring efficient and best utilization of available resources controlled by IoT and secured by blockchain. IoT system may connect myriad of nodes or machines of various kinds capable of communicating with each other to get services whenever required.

This edited book presents a deep insight for modelling, procuring, and building the smart city plan using IoT and a security framework using blockchain technology. The applications of the latest technologies like Wireless Multimedia Sensor Network in 5G environment will achieve the aims of smart city development. To plan and develop state-of-the-art infrastructure of a smart city, the major domains on which IoT will consider are individual networks (such as education, transport, healthcare, agriculture, energy, buildings, and business), connecting with proper security and management.

This book focuses on use of IoT and blockchain to track the unlawful activities of smart city residents at all levels through their day-to-day transparent recorded activities. The smart city is just the beginning, and there is always room for advancement in the use of sensors, algorithms for subsystems and systems, security solutions, interactive kiosks, communication technologies, embedded systems, deep learning based healthcare systems, agriculture, waste conversion into energy, wearable medical equipment, smart transportation that makes trips for people with disabilities more enjoyable, etc. The future plan for upgrading the system as per technological advancements will also be discussed. This book on smart cities gives the tools for effective planning, development, and conservation resources with the help of IoT sensors. The concept of smart cities focusses on maximal use of public transport to minimize the pollution to save the environment and travelling cost as well. The IoT sensors provide the real-time tracking of public transport to the citizens of smart cities.

Editors

Dr. Rajendra Kumar
Sharda University, Greater Noida, India

Dr. Vishal Jain
Sharda University, Greater Noida, India

Dr. Leong Wai Yie
INTI International University, Nilai, Kuala Lumpur, Malaysia

Dr. Sunantha Teyarachakul
California State University, California, United States

Acknowledgements

The book *Convergence of IoT, Blockchain, and Computational Intelligence in Smart Cities* is an edited collection of contribution of researchers from academia and industry. The editors thank all the contributors from different countries who worked hard to complete the book within the deadline. The editors also thank to reviewers of this edited collection for their rigorous peer review of the contents. The editors would also like to thank their universities for providing necessary help and support to complete this project. Last but not the least, the editors thank their friends and colleagues for their valuable guidance in the project.

Dr. Rajendra Kumar
Sharda University, Greater Noida, India

Dr. Vishal Jain
Sharda University, Greater Noida, India

Dr. Leong Wai Yie
INTI International University, Malaysia

Dr. Sunantha Teyarachakul
California State University, California, United States

1 5G Intelligent Transportation Systems for Smart Cities

Wai Yie Leong and Rajendra Kumar

1.1 INTRODUCTION

Using wireless connections, connected or networked vehicles are a novel smart transportation system that aims to increase both efficiency and traffic safety. Vehicle communication refers to wireless connections made by vehicles with infrastructure, other vehicles, and wireless devices. These connections are also called vehicle-to-infrastructure, vehicle-to-vehicle and customized-vehicle networks. In the past 10 years [1], many academics, industries (producers, telecommunications, automation manufacturers, etc.), governments (transport ministries), and professional standards associations (such as ASTM, IET, ISO, IMechE) have all demonstrated a great deal of interest in connected vehicle communication lines. The interplay of these parties has led to numerous research, innovations, and development projects, including Intellidrive, CVIS, and AHS. It has aspirations to develop linked car communication technologies and offer a proof-of-concept. The primary purpose behind the development and evolution of linked vehicle communication is mainly to maintain the security of the apps by arming the driver with the information and support they require to help prevent accidents on the road [1]. Wireless communications technology must consistently offer the appropriate quality, coverage, capacity, and availability in order to execute the quality of service (QoS) with respect to latency. This is due to the fact that scalable, reliable, and affordable core technology programmes are necessary for connected vehicle communication applications' public safety. The latency and reliability requirements for the safety applications may be met despite the possibility of many non-safety approaches and applications being enabled [1].

1.2 INTELLIGENT TRANSPORTATION SYSTEM

This system consists of complex and coordinated software and sensor-based information transfer applications, as well as the information addition process used by transport management. Intelligent transport uses advanced methods to efficiently manage traffic with information controllers [2]

1.2.1 Types of Intelligent Transportation Systems

The following five categories, which are shared by the parties involved, are used to categorize intelligent transportation systems (ITSs):

- Modern traffic control techniques.
- Modern portable information systems.
- Systems for commercial vehicle operations.
- Modern public transit infrastructure.
- Sophisticated vehicle safety and control systems [2].

The components of this system that are essential include:

1. **Information collection**: To assess the approaches at the level of strategic studies as development strategies for mass transit. It is therefore regarded as the priciest and most important component of such investigations. To use sensitivities in traffic management and to help lessen traffic congestion, this first phase comprises positioning them in specified locations on the highways. Additionally, it is connected to intelligent traffic signs, which gain from this knowledge. Second, traffic-related specialized cameras are employed to keep track of infractions, gauge the level of security, and examine certain aspects of the route [2].
2. **Data analysis and processing techniques**: These technologies comprise the software and tools used to turn raw data into information that drivers may use, such as reducing the speed limit or rerouting traffic based on the presence of traffic bottlenecks [2].
3. **Control and monitor methods:** To establish control rooms, information banks for traffic and transportation, traffic conditions, and warning signs. The methods also include coordination with public transit authorities based on various transportation matters [2].

1.3 DATA ANALYTICS ON VEHICLE CONNECTIONS

As indicated in Figure 1.1, the annual output of automotive vehicles is anticipated to achieve 800,000 units globally by 2030. This high level of automation can be employed to provide passenger transportation without the need for a human driver [3].

While we move closer to what many predict the future of driverless transportation will look like, as indicated in Figure 1.2, numerous businesses are actively engaged in the technology rivalry to construct a fully autonomous car. Driverless automobiles are already a reality thanks to the rapid advancement of artificial intelligence in recent years. In fact, on US public roadways, autonomous vehicles have reported travelling millions of miles, with very few terrible occurrences being reported in recent years [4].

Figure 1.3 compares information from two polls conducted by the National Safety Council. The first occurred in January, while the second occurred in late March or early April 2018. Self-driving cars were seen as less safe than human-driven autos

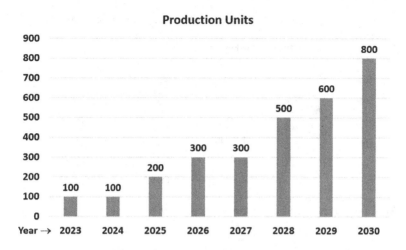

FIGURE 1.1 The expectation and estimation level of automotive production between 2023 and 2030.

by 36% of respondents in January. However, with the same worry index at the beginning of April, the replies had climbed by 14%. The public's interest in self-driving cars has increased as a result of some businesses, such Google spinoff Waymo (as shown in Figure 1.3) [5].

The United States is not yet prepared for autonomous vehicles. Because they will be safer than the alternative, everyone wants to possess a self-driving automobile. After all, Figure 1.4 depicts the annual number of car fatalities in the United States [6].

After two tragic crashes in the United States on April 16, 2018, the self-driving car is in the spotlight. Another incident included an autopilot on a special Tesla-based

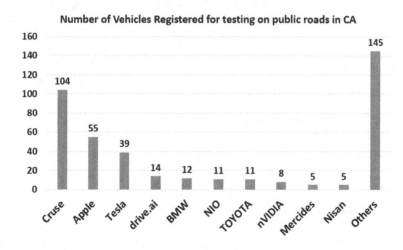

FIGURE 1.2 Statistics of autonomous vehicles recorded in California for testing on public roads.

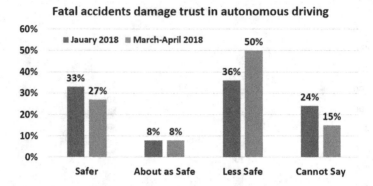

FIGURE 1.3 Fatal accidents undermine trust in self-driving vehicles.

Model-X which reported colliding with a highway divider. The first incident was an Uber autonomous car that also killed a pedestrian. Just before these two instances, Ipsos performed a significant poll on opinions regarding self-driving cars and discovered that the developing world had the highest level of support for them.

Self-driving cars are especially popular with Indian participants. Forty-nine per cent of the interviewees support them and eagerly want to use them. The 5% would never use it, while another 46% said they are dubious of them but found the notion interesting. Similar levels of optimism were seen in China, however, in Russia, only 33% were in favour and 59% were not sure.

Six out of ten Brazilian respondents in South America are still sceptical but inquisitive, while three out of ten stated they eagerly wanted a driver in a self-driving car.

FIGURE 1.4 A statistic on US consumers buy self-driving cars.

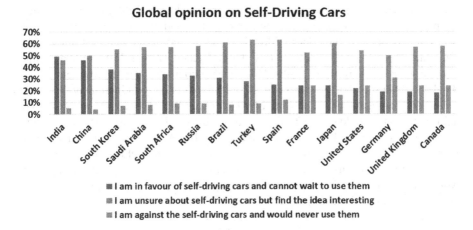

FIGURE 1.5 Divergent views on self-driving cars around the world.

Before the two instances, scepticism was already widespread in the United States. The 25% of those polled there indicated they would never employ self-driving technology, while 54% said they were unsure yet curious. Figure 1.5 shows that support for the driving exam was much lower than in India, at 22%, who also indicated they could not wait for the test [7].

1.4 EVOLUTION AND HISTORY

In the late 1990s, cars started to connect primarily with external sources for safety. The first versions of the autos had cellular connectivity for emergency notifications. Beginning in the new millennium, communications improved in terms of security and comfort, added remote features, and finally connected to cell phones [8]. A similar time frame saw the beginning of connected vehicle protection arrangements made by G + D Mobile Security. Since 2010, the extended network has made it possible to use stop and review function, self-driving mode, remote control tool, and advanced security driving features.

The primary highlights of the initial years, late 1990s, the initial automotive telematics benefits, were crash alerts to crisis responses, vehicle tracking, and emergency support. OnStar's fundamental cell structure was unveiled in 1996. The frameworks of the airbags connect with an OnStar advisor as they expand, providing information to emergency responders. Mercedes-Benz introduced the primary remote key in 1997 under the name "Keyless Go." With BMW's assistance, data administrations were advanced. In 1998, GM improved OnStar management by modernizing infrastructure and introducing vehicles with speech recognition and hands-free calling capabilities [8].

In 1999, TeleAid remote assistance with emergency aids and found automobiles was sent out by Mercedes-Benz. The Services promoted in this arrangement included advanced diagnostics, route, online connections, car-log, and mobile phone applications with limited online and remote features in the 2000s. A new Mercedes S-Benz specification and model that can remotely unlock the door is highlighted by TeleAid. With over 8 million connections, OnStar began providing continuing traffic data and remotely opening the door in 2001. The continental-driven demonstration tools for remote vehicle control designed by the GM Good Wrench Remote Diagnostic Service were made available in 2003. Through a computerized arrangement to get the device, Mainland offers car wellness and versatility gadgets incrementally.

In 2005, navigation was activated by carrying out each stage on its own. The administrations for slow vehicles were extracted from OnStar in 2007. The initial iPhone was launched. In the United States, the BMW Assist administration was launched with four years of free assistance remembered for vehicle purchases. The vast majority of the current IT units that are currently being transported around the world are a result of Mainland's entry into information-only technology, also known as "machine-to-machine data developments" [8]. The launch of a new Android cell phone occurred in 2008. Avoid Chrysler and Pocket Uconnect Web, which is powered by Autonet Mobile and offers easy 3G WiFi hotspots for constant web connection. The device must be installed by the vendor.

Mercedes-Benz introduced mbrace administration in 2009, allowing owners to communicate via iPhone and BlackBerry. This can be achieved by remotely locking a door, unlocking, and locating the vehicle. There were more car affiliations via various devices in 2010. The development of electric vehicles has done nothing except increase the demand for wireless and remote applications. A few cars have online programming updates. More connections and updated information are needed for new administrations. In V2I and V2X innovation, the primary vehicles are visible, especially for remote application and administration expansion in new structures. The CarWings iPhone app, which provides access to battery charging, car-driving range, and temperature monitoring, was used to power the Nissan LEAF. At CES 2010, the OnStar Mobile application was revealed.

The programme not only opens doors but also lets Chevy Volt owners specify charging times. In 2011, OnStar launched a new feature on FamilyLink, which has been used to locate family members, deliver email, and set instant messaging alarms. They also promote Nissan LEAF applications for smartphones running on Android and BlackBerry. The Mercedes-Benz mbrace has several features, including Facebook, Yelp, internet browser, geolocation, alerts, and an expanded feature to share location with friends in 2012. New York Auto Show, Google Earth route, 3G web association premiere for the Audi 7. The original Tesla Model S, which supports 3G and can update programming via online, was presented.

The Audi A3 was the top vehicles in Europe in 2013 for having a 4G LTE information association. Many of the GM 2015 vehicles in the United States can access WiFi 4G LTE hotspots as of 2014. Tesla S models come with devices that, after a future software update, enable autopilot features. The vulnerability in 2015 suggests that ADAC (Allgemeiner Deutscher Automobil-Club) analysts could remotely unlock cars as well as copy BMW servers. New and light vehicles have been reported by the European Parliament to naturally alert salvage organizations when a vehicle accident occurs in 2018.

It is known as e-Call. The driver can manage to exit the vehicle outside using the Mercedes-Benz Remote Parking Pilot programme. From a PC 16 kilometres far away,

5G Intelligent Transportation Systems for Smart Cities

Charlie Miller and Chris Falasic remotely hacked and intercepted a Cherokee jeep system. It regulates the wipers, radio, windshield, and quickening agent. Mercedes-Benz introduced the new E-Class system with the first creation model including Car-to-X compatibility. Tesla Motors released an online 7.2 software update for Autopilot that includes self-driving technology.

The Hacks discovered a number of flaws in various models in 2016, allowing programmers to take over auto vehicles and even drive them away. To control exiting from outside the vehicle, Tesla discharges 7.1 with the Summon highlight. For a few 2017 Audi vehicles, Audi America announced V2I innovation (traffic light data). BMW Connected Version 3.0 for iOS comes with Alexa, a first mile route, vehicle administration alarms, and an Apple Watch app. BMW Connected 1.0 for Android is accessible. In 2017, G + D Mobile Security joined BMW and its Connected Drive Services as a trusted e-SIM board partner [8].

1.5 HOW LINKED VEHICLES FUNCTION

There is still considerable study to be done and the accompanying vehicle framework is still under development. Committed short-range communication (DSRC), a technology similar to WiFi, will likely serve as the foundation for the security-related frameworks of linked vehicle innovation. DSRC is efficient, safe, reliable, and operates with a high degree of commitment. Unreliable applications may rely on several forms of distant innovation [9].

The ability to "speak" to one another within a car or post-retail gadget that is continually transmitting critical security and transportation information will be available to automobiles, trucks, transports, and other vehicles. Traffic signals, work zones, pay booths, school zones, and other infrastructure can all be communicated remotely by vehicles connected to the infrastructure [9].

Vehicles cannot be followed because of the mysterious nature of the vehicle data being supplied, and the framework is changing the safe. Associated cars provide drivers with a 360-degree awareness of similarly equipped vehicles within a range of about 300 metres. This secure framework doesn't track your car and hides personal information. Drivers will receive warnings via a visual screen presentation and alertness, driver seat vibration, or a special tone that informs them of potential threats. These are merely reminders to keep the driver in constant control of the vehicle.

However, the technology can also support limited robotized tasks when the driver only attempts to maintain partial control of the smart vehicle. Drivers can respond swiftly to alerts to prevent dangerous collisions. For instance, a warning is provided to drivers in the application for convergence development support when it is unsafe to approach a crossing point. When it is unsafe to pass a slower-moving car, the don't pass application warns drivers. The emergency electronic brake light application alerts the driver when a distant car several vehicles in front is slowing down and helps them anticipate the location of this susceptible side.

Admonition software that enables the company driver to essentially see what is happening in their most vulnerable area. Even if a motorist cannot see or hear an incoming train, connected automobiles can communicate with sensitive infrastructure, such as rail level intersections, to alert them to the impending arrival of a train. The associated cars can also assist with traffic situations and health challenges caused by the climate. This is especially true in situations like dark ice, where the weather may not appear dangerous but the streets are difficult to navigate. Information obtained from several nearby vehicles might help determine whether a risk might be present.

During the winter season, the Traffic Management Centres (TMCs) can receive road temperature data from the cars, providing granular ongoing data to help screen and oversee the execution of the transportation framework. The centres would then have the ability to carry out tasks including changing traffic signals and speed restrictions, informing maintenance crews, sending assistance cars to the road, and broadcasting warnings to drivers so they can get constant Road temperature data in intelligent transportation [10].

1.6 SMART CAR

The smart car, which is one of many high-tech integrated transportation firms and the primary technology and innovation of the intelligence transportation system, is reviewed in this study. It is a collective name for comprehensive automotive technology that can fully or partially perform one or more driving functions. In the past 20–25 years, a significant area for robotics applications has evolved which is focused on cars and smart vehicles [11]. One of the most significant inventions of the 20th century was the automobile. It created a sizable industry and allowed people to roam about freely, dramatically altering our way of life. Yes, the vehicle had a major role in the considerable transformation given the way our modern economies are set up.

Based on statistics, more than 800 million cars are found on the road and it is estimated that the number will rise during the following ten years. The establishment of several research initiatives that seek to develop the common actions that drivers carry out while driving has actually been prompted by this significant obstacle. The term "smart car" refers to a vehicle with operating systems that automate driving activities such as selecting the safe lane, avoiding obstacles, navigating through traffic jams, pursuing the car in front of you, assessing and avoiding dangerous situations, and choosing the lane. Improving driving efficiency, comfort, and safety has generally been the driving force behind the development of smart cars [11].

1.7 FUNDAMENTAL AUTONOMOUS VEHICLE TECHNOLOGY

The term "smart vehicle technology" typically refers to the advancement of autonomous vehicle functionality. The ability to perceive the environment and the state of the vehicle; the ability to communicate with the environment; and the ability to plan and carry out the most effective exercises are the major characteristics of smart vehicles.

The automobile industry, academia, and governmental organizations are all looking for experts in the quickly expanding field of smart vehicle technologies. The broad interest in smart car technology is also being stoked by a plethora of UGV competitions taking place all over the world [12].

Although self-driving vehicles are not yet on the radar of autonomous vehicle manufacturers, these technologies are swiftly innovating into passenger vehicles to help the car driver in life-or-death circumstances. The automobile industry's next generation of vehicle protection systems will make use of linked car technology. The "Smart vehicle systems" are described by Richard Bishop as technology developments that monitor traffic conditions and provide data or vehicle power to assist drivers in doing the best possible vehicle activity [12].

1.8 THE INTELLIGENT DRIVER MODEL AND APPROACH

The intelligent driver model (IDM) approach is the model that smart car simulations employ the most commonly. The IDM approach has no set reaction time and is provided in the form of a continuously differentiated acceleration function. It is more comparable to the features of ACC cars than to those of a human-driven vehicle [13]. The IDM can be utilized as ACC or a human-driven car by altering a few settings. Additionally, CAV simulations employ the modified IDM [14]. IDM's primary purpose is as follows:

$$a_{\text{IDM}}(s, v, \Delta v) = \frac{dv}{dt} = a\left[1 - \left(\frac{v}{v_0}\right)^\delta - \left(\frac{s^*(v, \Delta v)}{s}\right)^2\right] \quad (1.1)$$

$$s^*(v, \Delta v) = s_0 + vT + \frac{v\Delta v}{2\sqrt{ab}} \quad (1.2)$$

The IDM produces acceleration and deceleration rate in the majority of circumstances. When the present vehicle-distance is much smaller than the intended vehicle-distance, the deceleration rate becomes unreasonably high. When it comes to human-powered cars, with the most severe deceleration rate, the vehicle in front of them won't stop abruptly for no apparent reason. Therefore, the current difference is said to as a somewhat mild-critical condition [15] when it exceeds the optimal difference distance. Do, Rouhani, and Miranda-Moreno addressed this problem by combining the IDM with the Constant Acceleration Heuristics (CAH) to minimize the excessive deceleration speeds. This CAH model's fundamental assumption is that the prior vehicle wouldn't abruptly change the acceleration rate into seconds [16].

Based on study, three special conditions must be satisfied for the CAH to function properly: (i) there is never a need for a safe time ahead or minimum distance; (ii) the vehicle owners have to react immediately (with zero response time); and (iii) the acceleration of the vehicle under consideration and the preceding vehicle won't change in the applicable future (typically a few seconds) [17].

Numbers for the actual gaps, current vehicle speeds, previous vehicle speeds, and acceleration rate are presented. The maximum acceleration rate that avoids car crashes is provided as:

$$a_{\text{CAH}}(s, v, v_1, a_1) = \begin{cases} \dfrac{v^2 \tilde{a}_l}{v_1^2 - 2s\tilde{a}_l} & \text{if } v_1(v - v_1) \leq -2s\tilde{a}_l, \\ \tilde{a}_l - \dfrac{(v - v_1)^2 \theta(v - v_1)}{2s} & \text{otherwise,} \end{cases} \quad (1.3)$$

which effective acceleration rate = min (to prevent defects that might have been brought on by cars that came before it and were capable of accelerating more quickly). If the cars came to a complete stop when the minimal gap, $s = 0$, was reached, the assertion would be true. Otherwise, negative approach rates are eliminated by the Heaviside step-function, Q, because they are incomprehensible to the CAH. Leong and Ee [17] suggested the ACC model by combining IDM and CAH acceleration.

Based on the following factors, the ACC model generates various acceleration speeds depending on IDM or CAH. If both IDM and CAH arrive at the same speed result, the ACC model generates the same acceleration. The condition becomes slightly problematic if the IDM generates an unreasonably large deceleration while the CAH's deceleration is within a tolerable deceleration range. The ACC's acceleration is still greater than the CAH's acceleration after subtracting the deceleration in this situation. If either the IDM or the CAH yield acceleration that is significantly below b, there is a serious issue. The maximum combined acceleration of the IDM and CAH shall not exceed the ACC's acceleration. Acceleration of IDM and CAH should be a differentiable and continuous function of ACC acceleration [16].

$$a_{ACC} = \begin{cases} a_{IDM} & a_{IDM} \geq a_{CAH}, \\ (1-c)a_{IDM} + c\left[a_{CAH} + b\tanh\left(\frac{a_{IDM}-a_{CAH}}{b}\right)\right] & otherwise \end{cases} \quad (1.4)$$

The ACC model in Eq. (1.4) has one extra parameter, just like IDM. This extra parameter is called a concern for coolness. The ACC model relates to the IDM, at very small distances, the distance intensity varies but no velocity difference takes place. Al-Khateeb and Johari [18] predicted c to be equal to 0.99.

Using IDM as the benchmark model, the Cooperative Intelligent Demand Model (CIDM) was created [19]. They examined the functionality of CAVs in the system. The Human Driver Model (HDM) uses a spatial anticipation approach to achieve CAV communication.

The CIDM is subjected to the HDM anticipation, which divides the IDM into Eq. (1.5) based on Eq. (1.1).

$$a_n(\Delta x, v_n, \Delta v) = a_n^{free} + \sum_{m}^{n-1} a_{nm}^{int}(\Delta x_{nm}, v_n, \Delta v_{nm}) \quad (1.5)$$

The foundation IDM Eq. (1.1) is made up of two basic components: (i) Breakage, which compares the current distance to the expected, and (ii) Acceleration, which compares the present velocity with the desired velocity. When V2V activity is acknowledged in Eq. (1.5), the description is the same as in Eq. (1.5).

1.9 VEHICLES CONNECTIVITY AND NETWORKING

Networks can work in a variety of situations to stop collisions between specific cars. Vehicles that can speak with one another may be able to reduce individual car accidents or large-scale collisions [19]. To achieve the greatest outcomes, accident

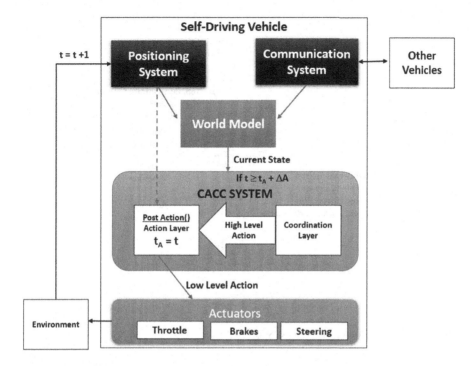

FIGURE 1.6 CACC layout.

avoidance may also utilize linked systems and autonomous independent systems. To enhance the performance of a potential dual system that combines autonomous and integrated systems, researchers have started to merge a self-control framework with machine learning techniques [20, 21]. Cooperative Adaptive Cruise Control (CACC) is an extension of Adaptive Cruise Control (ACC). CACC is successful in developing simulations, demonstrating the value of deploying autonomous and integrated CACC systems. The conceptual degradation of the CACC definition for readers is depicted in Figure 1.6. The GPS receives information from the atmosphere and filters it through it so that it can make the necessary adjustments. This can be expanded to a useful application with specific improvements to the infrastructure and technology [19].

1.10 ENABLING AND FACILITATING TECHNOLOGIES

1.10.1 ENVIRONMENTAL STATE

Sensing the environment around the car is a key component of smart car systems [11]. In smart automobiles, understanding the traffic situation is the hardest task to complete. It entails identifying the key elements, including the lane, other onsite vehicles, walking pedestrians, traffic conditions, road signs, and other ad hoc obstructions. Pace control when an event is discovered on a road scene is the most

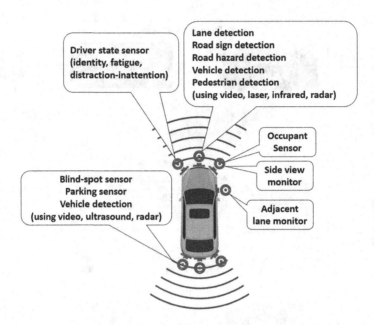

FIGURE 1.7 Environmental state sensing.

challenging issue. Ultraviolet, ultrasound, radar, laser rangefinders, and computer vision are popular sensors that scan the world as shown in Figure 1.7.

Infrared and ultrasound are often employed to detect impediments nearby, whereas radar is frequently used to spot barriers distantly. Frequency spectrum and image collection and analysis are employed to give additional information about the road scene in a variety of environmental conditions. Only a visual sensor can detect any elements of the road situations, such as traffic conditions and road signs. Sensor fusion is frequently employed in smart car applications, particularly in monovision and radar/laser sensors.

In partnership with major automakers, level one automotive suppliers Bosch, Denso, Delphi, Visteon, Siemens, and NEC mostly conduct this research. These sensors can be used to map the area surrounding the vehicle, which can then be used with methods like simultaneous localization and mapping [22] to create the intricate motions needed to park the car or avoid obstructions [11].

1.10.2 Tracking Lane

Since the initial implementations in the early 1980s, some car innovations are fitted with lane and obstacle identification and monitoring equipment. Therefore, computer vision is essential. Although laser scanners can also scan the path, computer vision and image processing techniques can accurately determine lane structure and lane markings. The majority of lane monitoring strategies are built around identifying lane signals and taking advantage of the local topography, such as comparing left and right lane signs, using a predetermined road width, or anticipating a flat course that is frequently traversed.

FIGURE 1.8 Lane detection.

These presumptions were primarily employed to address the problem of just offering one camera (a cost-effective option). Certain gadgets that use stereoscopic vision for track line identification may function without any limitations. The performance of the commercial lane tracker is found in Figure 1.8. Although lane recognition systems normally operate with 95% to 99% reliability, these systems cannot guarantee 100% reliability. Since autonomous driving cannot tolerate failures, lane detection devices are mostly utilized in lane deviation alert systems. Algorithms are being developed in an effort to achieve 100% driving efficiency in a variety of driving situations [11, 23].

1.10.3 Identifying and Detecting Traffic Signals

Identifying and recognizing the road sign is another application of machine vision, specifically for detecting traffic lights. Road signals are designed with individual users in mind. Clearly defined combinations of shapes, colours, and designs are used in road signs. At the right heights and places, banners are positioned along the route. Machine vision is also capable of interpreting traffic signs. The combination of detection systems and light detection allows for recognition and identification.

This task is often completed using pattern recognition approaches such crosslinking between photos and neural networks or vector support machines due to the constrained and well-defined possible range of road signs. Figure 1.9 illustrates the idea of a speed warning device based on velocity signal detection. The reliability and strength of mark recognition identification pose several challenging research ideas. Many automakers are currently developing new technology [24].

FIGURE 1.9 Identification of street signs for level warning broadcast.

1.10.4 Cars Detection

Attempts have been made to detect cars using a variety of sensor technologies, including sonar, lidar, radar, and vision. The substances' qualities are the same, regardless of their various shapes and hues, and they may be identified by their large size and translucent surface. Also, the vehicle's location can be anticipated till the tracking of the street direction is present. Several independent sensors may successfully identify vehicles automatically.

There is a technology that uses vision to identify cars. Although the answer to this question might seem simple, each sensor has a unique set of applications and difficulties. Although vision is often good, it can be difficult when there is poor lighting, low visibility (such as at night or in tunnels), or when there is a lot of traffic. As a result of the fact that car mufflers and tyres often exhibit raised temperatures and may thus be easily recognized in the image, infrared vision (thermal imaging) can identify cars with a high degree of certainty. However, stationary vehicles such as trailers and cars that have just moving are less noticeable since they are colder than moving vehicles. Although lidars are generally reliable, bad weather reduces their sensitivity.

Although inexpensive, radars can have biased internal measurements because of the presence of some nearby reflecting objects. Sonar is only usable for at most extremely brief periods of time. The purpose of the effort is to introduce a successful multi-sensor combination. Radar visibility incorporation is a rising remedy [21].

1.10.5 Adaptive Cruise Monitor and Control

The proposed ACC is the classic cruise control's functional expansion to keep a secure distance from the car in front of you. The smart car uses automatic cruise control to maintain a fixed speed when there are no vehicles in front of it. An automobile

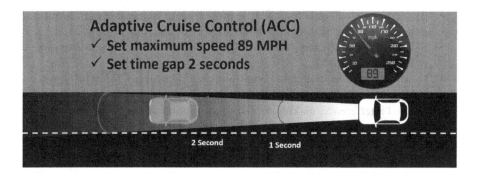

FIGURE 1.10 Adaptive cruise control (ACC).

with ACC may be failing to keep a safe distance after employing radar or lidar to detect a slower-moving car in front (range of 1.5–2 s).

An illustration of the ACC principle is shown in Figure 1.10. Because ACC-based devices are largely made for driving cars, the detecting duty is a little easier than the rear-end impact countermeasure difficulty. When using ACC, this issue is avoided because all stopped items are disabled. The biggest sensor challenge for rear-end collision-avoidance technology is differentiating stopped autos from off-road objects. Moving objects are categorized as in-lane or out-of-lane using a number of variables [25].

To confirm if the vehicles in front are on a given route, the smart car's steering range is employed as a street bend gauge. The products are specifically advertised as comfort products rather than safety systems, thus they are limited to handling routine scenarios with typically low-speed contrasts and leave the more challenging ones to the human driver. The driver is always on guard, keeping an eye on the road and taking in the surrounding traffic. Such initiatives are being introduced by both major automakers.

1.11 MODULATION SYSTEM FOR INTELLIGENT CARS

Originally created for RF communication, spatial modulation (SM) is a hybrid MIMO and modulation technology that enhances spectrum network performance while concurrently lowering device variability. Smart cars made use of SM-based V2V and V2I communication systems.

1.12 SPATIAL MODULATION (SM) ADVANTAGES AND DOWNSIDES

Following are the benefits of SM:

- SM completely avoids both ICI and IAS [26].
- SM only requires one RF chain on the transmitter [26]. The SM system of motion is to blame for this. Data processing is carried out by a single transmitting antenna while all other antennas are silent.

- By increasing the logarithmic field in proportion to the number of transmission antennas, the 3D diagram in SM shows a double spatial gain [27].
- SM provides an efficient, spectrum-based symbol with a greater than one equal symbol rate [27].
- In contrast to conventional spatial multiplex approaches for MIMO systems, interference de-mining algorithms are not required to resolve inter-channel interference, making receiver construction simple [26].
- SM will use a single current receiver to access ML decoding [26].
- SM will work well if the Nr < Nt.
- SM can function in a variety of control scenarios. For the reason that distinct sets of transmitters and receivers, it frequently takes up a variety of spatial places. Other users can access the same wireless services when all receivers use the channel pulse array replies of all transmitters to detect data (for instance, many users are discovered) [27].
- SM is more powerful than traditional low-complexity encryption methods in MIMO systems [26].

Following are the disadvantages:

- To use the SM principle, at least two transmission antennas are required [27].
- SM model might not be employed or produce satisfactory performance if the radio's wireless transmitter and receiver links are not sufficiently separate. This limitation is quite comparable to conventional spatial multiplexing methods, which allow for a rich dispersion setting to greatly improve the rate of change [27].
- The channel estimator may be subject to certain limitations. The channel receiver requires awareness of the whole channel to recognize data [26].
- SM only causes the data rate to increase logarithmically (as opposed to linearly) with the transmitting antenna level. SM may make it more difficult to achieve extremely high spectrum efficiency for a practical number of transmitter antennas [26].

1.13 THE DEVELOPMENT OF INDUSTRY REVOLUTION 5.0

Smart mobility solutions are greatly needed due to the significant costs associated with urban traffic congestion, but as self-driving cars become more prevalent, their demand will only increase. In order to enable and manage fleets of automated vehicles at the city level, intelligent mobility services still need to develop and combine technologies like cloud infrastructure, 5G, the Internet of Things (IoT), smart sensors, and artificial intelligence, self-driving cars would be heavily reliant on IT networks [28].

As a result, within ten years, smart transit networks will serve as the central nervous system to develop sustainable cities. Examples from Toronto and Lisbon provide us with a sense of how such networks operate. Although they would have a big impact on developing communities, smart transit networks would also play a significant role in changing culture. As long as ITSs are widespread, numerous

5G Intelligent Transportation Systems for Smart Cities

industries will change as well. To start, a scientist with fully autonomous vehicles would not require the majority of insurance types.

Infrastructure is currently out of control, which will significantly modify real estate prices and push police and other public security agencies to divert resources from car-related issues (such as traffic fines, which may disappear). You need a strong information and communication infrastructure, and to acquire, process, retain, and analyse these data, breakthroughs like 5G communications, smart sensor networks, big data, artificial intelligence, and cloud computing are needed [23, 24, 29].

1.14 SIGNIFICANT TECHNOLOGIES IN INTELLIGENT TRANSPORTATION SYSTEMS

1. Advanced and improved tracking system: The majority of cars today come equipped with GPS systems. The GPS provides two-path communication to aid traffic officials in finding automobiles, monitoring vehicle driving speed, and providing critical services. The cell phones, multipurpose apps, and Google navigation maps have evolved into useful tools for tracking them down, learning about the quality of the streets and the volume of traffic, and locating different positions and locations [30].
2. Smart sensing technologies: These technologies equipped both automobiles and street systems with smart sensors. The radio frequency identification (RFI) and sensor architectures guarantee walkers' safety. The street reflectors and inductive circles are widely dispersed, these will help with traffic management and safe driving, especially at night. Additionally, they can tell how much traffic there will be in a certain period of time and can distinguish between fast and slow-moving automobiles [30].
3. Advanced vehicle detection: CCTV or external cameras would help traffic chiefs with a number of concerns. Administrators may find it easier to monitor traffic flows, spot crisis situations, and identify road blockages if key locations and important crossroads are captured on video. Robotic number plate recognition is integrated with car controls, allowing for the tracking of vehicles for safety reasons [30].
4. Advanced traffic light system: The congestion light system during rush hour often uses radio frequency recognition (RFID). When used to a few streets, street intersections, and autos, innovation provides the accurate calculation and database. These lights might be altered without a person's intervention during peak and important traffic hours [30].
5. Immediate emergency E-call vehicle service: During an emergency, such as an accident or other tragedy, the in-car sensors can contact the nearby crisis jog. The e-call can let the driver to communicate with the certified administrator and will also properly transmit important information to the jog, including the date, location, position of the vehicle, and vehicle description. In all of Europe, e-call administration is now required for newly registered automobiles [30].
6. Precise route information: If drivers know which route is best for their vacation, it will be easy and relaxing for them, especially if they are traveling to

a different district. Travel can be made easier with constant information on traffic patterns, route frameworks, sharp-turns, stopping signs, and street conditions. Before starting the trip, the driver can obtain this information through their screen, portable device, or online media communications and networks. Additionally, they can monitor drive duration, traffic stream level, elective roadways, street construction projects, transportation methods, cost rates, and halting offices using correspondence, remote devices, and variable message signs (VMS) [30].

7. Safety and vehicle control: Through alert and vehicle control nuances, the administration provides vehicle administrators with well-being assistance. Drivers will evaluate their driving skills, the quality of the roads, and the efficiency of their vehicles. By studying the paths of various vehicles, they will be aware of any front or backside accidents as they change directions or transfer lanes at intersections. When visibility is poor due to extreme weather or poor night vision, propelled cameras in automobiles can regularly assist drivers by capturing pictures of the distinctive area. The goal is to warn drivers and rescue vehicle workers of impending accidents in order to prevent accidents or episodes [30].

1.15 SMART TRANSPORTATION STRUCTURES AND ARCHITECTURES

The fundamental tenet of smart transportation is that individuals should have access to the proper information at the proper time, place, and computer to enable them to confidently make transportation decisions and to make it simple for them to employ better, quicker methods. In order to accomplish this, we require software architecture. These systems handle a variety of tasks, including parking, traffic, exchanging data, public wide transportation, network routing, and energy consumption management.

1. Graphical-oriented transportation using information from the IoT. At every intersection of the street, street sensors are installed to measure the traffic details on the road, such as the number of vehicles, average vehicle speed, traffic volume, size, and so on. While the vehicle configuration is used to gather information on the specific vehicle, such as location and speed. They connected the information from the two frameworks to the database via hand-off hubs, networks, gateways, and classifications.

    The primary aspect mentioned in this Smart Transportation Architecture is used to identify the most efficient route, from point A to point B depending on the traffic conditions currently in place, or it could be used to alert traffic experts to things like street blockages, increased traffic and injuries [31].

2. Transportation System Architecture for a Multi-Station Shared Vehicle System: In reality, a community automobile network is made up of several vehicles that are used often throughout the day by numerous individuals [31]. An example of multi-station shared vehicle system and framework is presented in Figure 1.11.

5G Intelligent Transportation Systems for Smart Cities

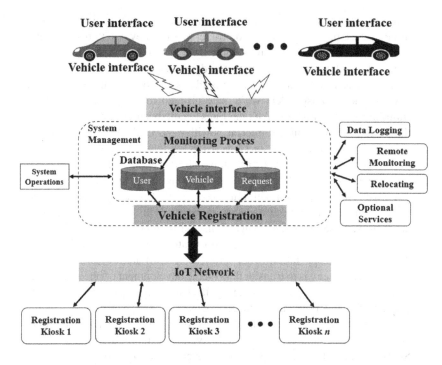

FIGURE 1.11 Multi-station shared vehicle system and framework.

This architecture is made up of three main parts:
a. The trip registration component for users
b. The element of system management
c. the parts and components of the vehicle.

The user trip-registration component's main objective is to enrol a vehicle application that needs enlisting (bottom section of the illustration). The customer must have a contactless card for each users and a touchscreen display in order to request a vehicle. The next section, called system management component. This system is based on a running database with information on customers, vehicles, and networks.

Smart agents are always updating this database. The registration process, monitoring process, system operator interface, data logging, and additional remote processing are only a few of these sharp experts. An electronic enrolment procedure is in charge of monitoring data from access to the gateway. The fundamental is the observation component, which continuously monitors data coming from the vehicle armada, sends "registration" signs to the designated vehicle, and routinely measures travel duration and separation toward the end of the excursion. The framework administrator follows each vehicle's condition in a guide-based manner thanks to the gadget administrator interface.

Both vehicle subtleties and registration booths may be focused to learn more about customer driving patterns, vehicle behaviour, and the reliability of the executives' processes. Remote controls allow for programmed control of device movement, charging customers as they arrive at predetermined times, pursuing vehicles to protect them if they don't cross predetermined cutoff points, recording vehicle trouble warnings, recording vehicle time of arrival, and anticipating traffic demands.

The UCR IntelliShare vehicle module is the third component of this device. Radio transponders were once utilized to communicate with the framework's administration. It integrates the method of reasoning that forms the basis of convention and interfaces with parts of the vehicle control signals, such as door lock and start enact/stop hardware. In order to determine the battery's charge status, the assistant battery voltage, odometer beat estimates, entrance open/closed signs, charging sign, and sign scope of the rigging, work in cooperation with multiple signs on the car. This framework's main element is routine armada maintenance. This method will improve air quality, lessen stopping congestion, and increase the allure of surface vehicles. Both clients and device administrators feel the pinch [31].

3. The T-CPS (transportation-cyber physical system) architecture. The main goal of this strategy is to transfer data on the development of physical objects and the state of the system to the digital framework, then connect digital frameworks and physical segments focused on the assembly of calculation, interchanges, and control advances to achieve data exchange, device synchronization, and efficient vehicle dynamics. Its model implements an SOA-focused collaborative programming construction philosophy [31].

The perception layer is the top layer in this system design. This layer makes use of sensor and collection hubs. The ability to observe any physical resource in the world, including cars and utilities, is crucial. The networking level, includes all physical stations and hubs. The function is mainly to ensure contact productivity while transferring the initial piece of information that the upper layer has decoded to the data centre. The computational bit is the third part.

This layer's main task is to maintain proficiency in continual information assessment and reenactment. A layer of impact makes up the fourth layer of this structure. This device can provide a science control calculation based on traffic mass information, issue guidance to the control hubs, or convey a crisis readiness signal through the actuators. The operation sheet comes next. With a constant traffic data office, its software would flexible the customer terminals.

No fresh component that tends to be taken into account by the general vehicle arrangement is investigated in this system. This system connects actual physical components to the digital web, but it has some drawbacks, including maintaining very high availability, ensuring the dependability and consistency of this framework to achieve adequate control effectiveness, adapting suitably to irregularities and delivery, and enhancing the accuracy and usefulness of the conveyance of traffic data [31].

1.16 APPLICATIONS FOR INTELLIGENT TRANSPORTATION

1. Electronic toll collection (ETC) reduces gridlock at toll booths and automates toll collection by allowing vehicles to pass through toll gates at the rate of traffic. With the use of ETC lanes and cordon zones in city centres,

congestion pricing is now being enforced using more advanced technology. ETC systems were initially employed to facilitate toll fee collection. Up until recently, the majority of ETC systems were focused on radio technology in automobiles to categorize the vehicles as they passed by the gantry using patented protocols [32].

2. Dynamic traffic light sequencing: Bypassing or avoiding issues that typically occur with systems that make use of image processing and beam interruption techniques, dynamic traffic light sequencing makes use of RFID. In order to give the best time-management strategy, a region with many vehicles, lanes, and roads is equipped with RFID system associated with a potent programmed algorithm and a huge database. Each board's passage is subject to a challenging timetable that is developed. In the presence of some extreme instances, the dynamic sequence algorithm adapts. The number of vehicles and the features of the routing can mimic a traffic police officer's decision [18]. An illustration of multilane traffic sequence is shown in Figure 1.12.

3. System for alerting emergency vehicles: An in-vehicle e-call is generated manually by the occupant(s) of the vehicle, automatically by the activation of the in-vehicle sensors following an accident, or both. The electronic call sends voice and data simultaneously to the closest emergency location (usually the public safety response point). The voice contact enables the

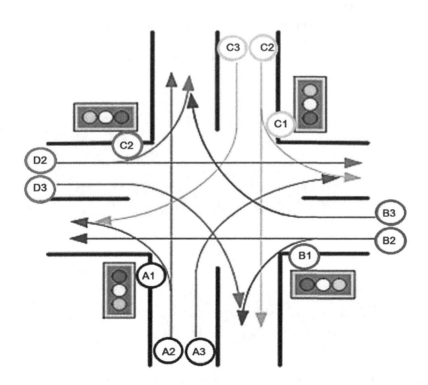

FIGURE 1.12 Flow of multilane traffic sequence.

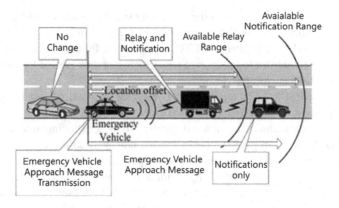

FIGURE 1.13 Information about emergency vehicles' approach.

occupants of the vehicle(s) to speak with the qualified e-call operator while also providing the e-call operator with the minimal amount of information necessary for them to take the necessary action, such as the vehicle's identity, the details of the accident, the date of the incident, the precise location, and the path of travel. Such scenario is presented in Figure 1.13. E-call is a popular option for all legal types of automobiles, primarily in industrialized countries [33].

4. Automatic road enforcement: A system for detecting and identifying vehicles that violate the speed limit or other laws on the road is shown in Figure 1.14. Following that, the defendant is accused based on the number

FIGURE 1.14 Allocation of ASE camera displayed at Dubai Smart City.

tag [34]. The owners are then sent traffic passes through email. The programme consists of:
- Speed cameras to detect cars driving too fast. Vehicle speed is also measured in various systems using electromagnetic or radar loops that are buried in each lane.
- Cameras at traffic light intersections that record when a car crosses a stop line while a red light is on.
- Cameras installed in bus lanes to detect vehicles utilizing bus lanes.
- Devices that detect vehicles crossing double white lines.
- High-occupancy vehicle (HOV) lanes use cameras to catch vehicles infringing HOV rules [34].

Collision-avoidance systems: Every vehicle has collision-avoidance technology that has been shown to lower accidents. Given that automakers are beginning to use front-facing sensors to aid with vehicle detection. All data are collected by the collision-avoidance system in order to identify any potential impediments.

The system may be able to complete tasks that lead to safe driving behaviour if the speed differential between the car and any other object is not too high. The ideal collision-avoidance system would merely issue a warning, allowing the driver to choose whether to brake or move away from the barrier [35].

1.17 CONCLUSIONS

Smart technology, such as in-car connectivity and sensors, will progressively gain more public awareness. The sole sign that robot technology is widely used in the automotive industry. This approach shows the industry's vision to make advances in total protection, placing attention on having secure systems. The security cost per vehicle in these businesses could be much higher than that of a vehicle. The working circumstances are radically different because specialists operate and handle demand. Governments will continually attempt to resolve obligation difficulties as producers gradually get the ability to generate products of unwavering quality and cost. Therefore, drivers are placed in patterns by frameworks [36].

REFERENCES

1. K. Jadaan, S. Zeater, and Y. Abukhalil, "Connected vehicles: An innovative transport technology," *Procedia Eng.*, vol. 187, pp. 641–648, 2017. https://doi.org/10.1016/j.proeng.2017.04.425
2. H. A. Idan, "Towards the Use of Smart Transportation System in Iraq," Faculty of engineering, University of Kufa, 2018. http://eng.uokufa.edu.iq/archives/5061 (accessed May 03, 2020).
3. I. Wagner, "Projected Production of Robo-Cars Worldwide 2030 | Statista," *Statista*, 2020. https://www.statista.com/statistics/1068697/projected-robo-car-production/ (accessed May 03, 2020).
4. F. Richter, "Chart: These Companies Are Testing Self-Driving Cars in California | Statista," *Statista*, 2018. https://www.statista.com/chart/13868/registered-autonomous-vehicles-to-be-tested-in-california/ (accessed May 03, 2020).

5. "Public Worries About Self-Driving Cars Are Spiking After a Pedestrian Was Killed by a Robo-Car I Business Insider," *Business Insider Australia*, 2018. https://www.business insider.com.au/public-concern-increase-safety-self-driving-vehicles-charts-2018-4 (accessed May 03, 2020).
6. Tehrani.com – Comm & Tech Blog, "Yes, Consumers Will Buy Self-Driving Cars – Tehrani.com – Comm & Tech Blog," 2015. https://blog.tmcnet.com/blog/rich-tehrani/technology/yes-consumers-will-buy-self-driving-cars.html (accessed May 03, 2020).
7. N. McCarthy, "Chart: Global Opinion Divided On Self-Driving Cars I Statista," 2018. https://www.statista.com/chart/13531/global-opinion-divided-on-self-driving-cars/?nr_email_referer=1&utm_source=Sailthru&utm_medium=email&utm_content=COTD&utm_campaign=Post Blast%28sai%29: Roughly half of all Americans expect self-drivi (accessed May 03, 2020).
8. "A Brief History of Car Connections I G+D Spotlight," *G+D Mobile Security smart*, 2017. https://www.gi-de.com/en/spotlight (accessed May 03, 2020).
9. United States Department of Transportation, "Intelligent Transportation Systems – How Connected Vehicles Work." https://www.its.dot.gov/factsheets/connected_vehicles_work.htm (accessed May 03, 2020).
10. United States Department of Transportation, "Intelligent Transportation Systems – Connected Vehicle Basics." https://www.its.dot.gov/cv_basics/cv_basics_benefits.htm (accessed May 03, 2020).
11. A. Broggi, A. Zelinsky, M. Parent, and C. E. Thorpe, "Intelligent vehicle," *Springer Handbook of Robotics* (pp. 1175–1198), edited by Bruno Siciliano and Oussama Khatib. Springer, 2008. DOI: 10.1007/978-3-540-30301-5_52
12. O. Gusikhin, D. Filev, and N. Rychtyckyj, "Intelligent vehicle systems: Applications and new trends," *Lect. Notes Electr. Eng.*, vol. 15, pp. 3–14, 2008. https://doi.org/10.1007/978-3-540-79142-3_1
13. W. Y. Leong, "Angle-of-arrival estimation: Beamformer-based smart antennas," *2008 3rd IEEE Conference on Industrial Electronics and Applications*, pp. 1593–1598, 2008.
14. W. Leong, J. Homer, and D. P. Mandic, "An implementation of nonlinear multiuser detection in Rayleigh fading channel," *EURASIP J. Wirel. Commun. Netw.*, vol. 045647, pp. 1–9, 2006, https://doi.org/10.1155/WCN/2006/45647.
15. W. Y. Leong, J. H. Chuah, and T. B. Tuan (Eds.), "*The Nine Pillars of Technologies for Industry 4.0*," Institution of Engineering and Technology, 2020.
16. W. Do, O. M. Rouhani, and L. Miranda-Moreno, "Simulation-based connected and automated vehicle models on highway sections: A literature review," *J. Adv. Transp.*, vol. 2019, 2019. https://doi.org/10.1155/2019/9343705
17. W. Y. Leong, and J. Ee, "A warehouse management system for 3 dimensional tracking and positioning," *Applied Mechanics and Materials*, Trans Tech Publications, Ltd, vols. 152–154, pp. 1685–1690, 2012. https://doi.org/10.4028/www.scientific.net/amm.152-154.1685
18. K. Al-Khateeb, and J. A. Y. Johari, "Intelligent dynamic traffic light sequence using RFID," *Proc. Int. Conf. Comput. Commun. Eng. 2008, ICCCE08 Glob. Links Hum. Dev.*, no. December, pp. 1367–1372, 2008. https://doi.org/10.1109/ICCCE.2008.4580829
19. D. Elliott, W. Keen, and L. Miao, "Recent advances in connected and automated vehicles," *J. Traffic Transp. Eng.*, vol. 6, no. 2, pp. 109–131, 2019. https://doi.org/10.1016/j.jtte.2018.09.005
20. W.Y. Leong, "*Human Machine Collaboration and Interaction for Smart Manufacturing: Automation, Robotics, Sensing, Artificial Intelligence, 5G, IoTs and Blockchain*," Institution of Engineering and Technology, Stevenage, United Kingdom, 2022.

21. R. Kumar, R. C. Singh, and R. Khokher, "Framework for modeling, procuring, and building systems for smart city scenarios using blockchain technology and IoT," *The Data-Driven Blockchain Ecosystem*, CRC Press, pp. 30–50, 2022. https://doi.org/10.1201/9781003269281-3
22. W. Y. Leong, and J. Homer, Blind Multiuser Receiver in Rayleigh Fading Channel, 2005 Australian Communications Theory Workshop, pp. 155–161, 2005.
23. W. Y. Leong, *"Implementing Blind Source Separation in Signal Processing and Telecommunications,"* Thesis, The University of Queensland, Australia, 2005.
24. W. Y. Leong, and D. P. Mandic, "Towards adaptive blind extraction of post-nonlinearly mixed signals," *2006 16th IEEE Signal Processing Society Workshop on Machine Learning for Signal Processing, 2006*, pp. 91–96, 2006. https://doi.org/10.1109/MLSP.2006.275528
25. A. Kesting, M. Treiber, M. Schönhof, and D. Helbing, "Adaptive cruise control design for active congestion avoidance," *Transp. Res. Part C Emerg. Technol.*, vol. 16, no. 6, pp. 668–683, 2008. https://doi.org/10.1016/j.trc.2007.12.004
26. Y. Bian, M. Wen, X. Cheng, H. V. Poor, and B. Jiao, "A differential scheme for spatial modulation," *2013 IEEE Global Communications Conference (GLOBECOM)*, Atlanta, GA, USA, 2013, pp. 3925–3930, doi: 10.1109/GLOCOM.2013.6831686.
27. M. Di Renzo, H. Haas, A. Ghrayeb, S. Sugiura, and L. Hanzo, "Spatial modulation for generalized MIMO: Challenges, opportunities, and implementation," *Proc. IEEE*, vol. 102, no. 1, pp. 56–103, 2014. https://doi.org/10.1109/JPROC.2013.2287851
28. M. Kuga, "Simultaneous," *Sen'i Gakkaishi*, vol. 40, no. 4–5, pp. P393–P395, 1984. https://doi.org/10.2115/fiber.40.4-5_p393
29. https://www.nec.com/, "How Will Smart Transportation Systems Work in Ten Years?," *NEC/orchestrating a brighter world*, 2020. https://www.nec.com/en/global/insights/article/2020022504/index.html (accessed June 10, 2020).
30. Smartcity, "How Smart Cities can Benefit from Intelligent Transportation Network," 2017. https://www.smartcity.press/intelligent-transportation-system-for-smart-cities/ (accessed June 09, 2020).
31. P. Parmar, and T. Champaneria, "Study and comparison of transportation system architectures for smart city," *Proc. Int. Conf. IoT Soc. Mobile, Anal. Cloud, I-SMAC 2017*, pp. 675–680, 2017. https://doi.org/10.1109/I-SMAC.2017.8058264
32. K. Sampoornam, "Electronic Toll Collection and Gate Automation," No. January, 2019.
33. S. Drakatos, N. Pissinou, K. Makki, and C. Douligeris, "A future location-prediction replacement strategy for mobile computing environments," *IEEE Wirel. Commun. Netw. Conf. WCNC*, vol. 4, April, pp. 2252–2260, 2006. https://doi.org/10.1109/WCNC.2005.1424866
34. F. Khairuddin, M. Hamzah, M. Yusof, and C. Ng, "The automated speed enforcement system – A case study in Putrajaya," *J. East. Asia Soc. Transp. Stud.*, vol. 10, February 2019, pp. 2133–2146, 2013.
35. Roboauto, "How it works? Collision Avoidance System - Roboauto - Blog," 2017. https://blog.roboauto.cz/how-it-works-collision-avoidance-system-d05bd9807f1a (accessed June 10, 2020).
36. A. Eskandarian, *"Handbook of Intelligent Vehicles"*, Volume 1, pp. 1–1599, Springer, 2012. https://doi.org/10.1007/978-0-85729-085-4

2 IoT Smart Environment
A Comprehensive Study on Challenges and Solutions

M. R. Sundarakumar, S. Sankar,
Somula Ramasubbareddy, Velleangiri Vinodhini,
and Balusamy Balamurugan

2.1 INTRODUCTION

Embedded technologies like computer communications, telecommunication networks, and mobile computing are rapidly developed in history and their main focus is to make global perception. The reliable data transfer and processing of information between the machine-machine vs. man-machine are the main aspects of communication (Alcaraz, 2019; Sennan et al., 2021). With this approach, the entire world has started research on an innovative framework to succeed in social needs. The growth of new innovative techniques has given different ideas and solutions for that but has some limitations. Finally, computer networks came into the digital world applications it has reached a major percentage of researchers' thoughts (Samanta et al., 2021; Sankar et al., 2021). Even though it has a few issues and challenges to overcoming real-time data transmission, it supports only permitted conditions. The entire world is expecting to access machines and other components as things that are connected to the internet. Hopefully, the new framework works with all the limitations and expectations namely Internet of Things (IoT) has taken a vital role in all leading technologies (Alotaibi and Elrefaei, 2018; Karthiprem et al., 2015; Sennan et al., 2021). IoT has successfully controlled and monitored a lot of machines with its features effectively. Nevertheless, it has a lot of issues with data transfer and control in several real-world applications. Especially the automation process of optimizing devices using IoT gives accurate results for machines. In this digital era, technology to transfer data has increased rapidly with high-speed internet connection connected over the network. But the other factors like software, sensors, actuators, and electronic components are essential things of the IoT environment (Alotaibi and Elrefaei, 2018; Mishra et al., 2020; Samanta et al., 2020; Sankar et al., 2020a). Different processes are done on the IoT platform for accessing various environments. The entire process of IoT is shown in Figure 2.1.

2.2 IoT CHALLENGES

The human role is very important in the IoT framework for controlling and accessing the machine's parts. But IoT offers a lot of benefits for humans to access real-world applications (Dabbagh and Rayes, 2019; Sankar et al., 2020b). While coming to the

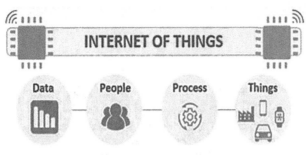

FIGURE 2.1 IoT process.

security and privacy concern the entire framework is making it difficult for accessing the devices. The major problem of IoT reveals in the process of handling huge data, integrity, security, automation, and monitoring with encryption capabilities. In this digital world transmitting text, audio, and video messages from one place to another place using an internet facility is complicated. Though the data size and its speed limit at a certain distance, some technologies are used to carry data to a long distance. The main problem of carrying data to a longer distance is secured transactions and timely communication (Dai et al., 2019). But the connectivity of machines has created certain issues in the IoT framework. A model has to follow for the secured transaction of data between the machines with some algorithms and methods in recent applications. The data dissemination among the machines produces challenges over the entire network communication. The operations performed in the IoT framework given their results based on the interoperability among machines (Ferrag et al., 2018; Sennan et al., 2020; Sankar and Srinivasan, 2018a). Hopefully, IoT has given solutions to most of the real-world applications. But some issues have to resolve with new technical approaches to satisfy social problems. The awareness of IoT technologies used in social problems emphasizes its features to the people who are accessing this framework. So, the above-said challenges are faced in IoT framework and it is presented in Figure 2.2.

2.3 IoT CHARACTERISTICS

IoT framework has its characteristics and this will be classified based on the behavior of the model used in its architecture. Most of the information transmits to a centralized controller and is typically monitored by network hackers using anomaly detection. Wireless Sensor Network (WSN) is an important technique used to

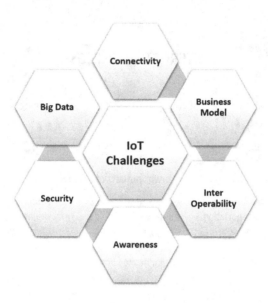

FIGURE 2.2 IoT challenges.

interconnect (Frustaci et al., 2017; Mishra et al., 2020; Sankar and Srinivasan, 2018b) the devices with its protocols. IoT framework is connected to the internet and WSN for accessing the devices, sensors, and actuators over a network. Data transmission occurs between the devices by WSN and carries the data to other places using the internet. Worldwide expansion can be done by WSN and the internet in IoT is a typical challenge for data transmission. Trillions of devices are connected over the IoT network between 2020 and 2025 (Ghafir et al., 2018; Sankar and Srinivasan, 2020). It requires high-quality techniques for connectivity and high-speed internet connectivity with WSN. It will be difficult to analyze the data that have been generated by sensors and humans. For the data analytics process to identify the exact device and their relevant data is a challenging role from a large number of devices present in the network. So, automation is needed to optimize the devices among IoT networks and vulnerability issues can be addressed in the future for a strong security framework (Habib et al., 2018; Sankar et al., 2020c).

IoT ecosystem also plays a major role in this framework and it will be used to connect the devices in various environments. Machine to Machine (M2M) is the concept of IoT (Hakak et al., 2019; Sankar and Srinivasan, 2018c) used to connect trillions of devices from various places but connected through routing protocols on IoT. Routing is the network concept used in IoT to connect devices remotely in various places (Hakak et al., 2018). For that several protocols are supported in IoT for data transmission. Connecting devices remotely used in IoT create issues in connectivity and data transmission. The overall concept of IoT environment characteristics denoted the challenges that have occurred in the ecosystem (Jindal et al., 2018) (Kamaruddin et al., 2018; Sankar et al., 2020a) for data transmission by sensors/devices in remote places. The solutions will be provided to these issues with recent trends and technologies in real-world objects. The solutions mainly deal with IoT characteristics like

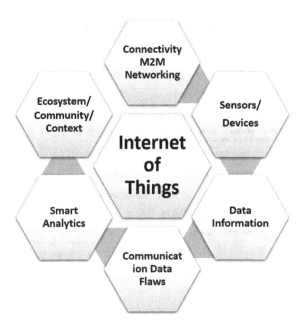

FIGURE 2.3 IoT characteristics.

speed, distance, connectivity, and transfer protocols (Khadam et al., 2020; Sennan et al., 2020). Figure 2.3 illustrates the characteristics of the IoT and their work.

2.4 IoT PROCESS MANAGEMENT

The process of the IoT can be classified with the help of their devices which are collected the data and analyzed them through proper algorithms. The data created by various sensors and devices are collected from different places but their size and other factors are varied in nature (Khadam et al., 2019; Sankar and Srinivasan, 2017; Vinodhini et al., 2020). To overcome the problem of this analysis has been made among those data and will be monitored at periodic intervals. Figure 2.4 depicts the detailed analysis of IoT environment processing methods.

2.5 IoT CHALLENGES AND ITS SOLUTIONS

There are multiple trends and solutions used to overcome the challenges in the IoT framework. Most of the challenges are concentrated on security side issues and privacy part problems. But the challenges faced in the IoT environment will make changes in futuristic approaches to the social problem needs (Kiani, 2018; Sankar and Srinivasan, 2016; Sennan et al., 2019; Sennan et al., 2020; Vinodhini et al., 2020). The following trends have been used to provide the solution for IoT challenges.

    a. Analytics at edge computing
    b. IoT security issues

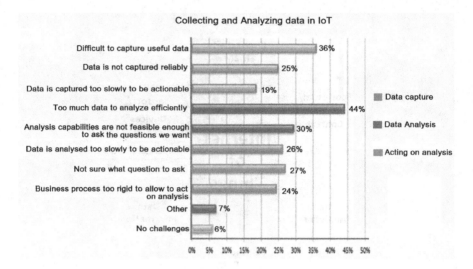

FIGURE 2.4 IoT process management.

  c. Industrial IoT challenges
  d. IoT cloud computing platforms
  e. ML/AI-based IoT applications
  f. IoT in sustainability and climate change
  g. IoT with block chain
  h. Cloud with edge computing in IoT
  i. Commodity sensors
  j. LPWAN and 5G network

2.5.1 Analytics of Edge Computing

When the sensors and machines are generated the day-to-day life data with real-world applications will be stored in a repository rapidly. After the successful storage completion, it might take analytics for data processing to lead the level of the machines to the next state. But the analytics part is quite complicated due to the size and time of the data generated (Sankar and Srinivasan, 2019; Sankar and Srinivasan, 2018d). So, it is used real-time data analytics on the network to stimulate the process in a high-speed task. In this stage, system security, latency, and cost are the major factors that decide the challenges. At last, the new term called edge is coming to the solution as local network devices to give the status of the machines on time. Some of the edge computing service providers are AWS, Microsoft Azure, and IBM cloud (Lin et al., 2017; Sankar and Srinivasan, 2016). They provide extended services to the IoT applications at their edge level. The companies that support this edge computing make their computers with gateways for easy transmission and control of data over the network. Edge computing promising technology working with machine learning (ML) and artificial intelligence (AI) techniques to perform data transmission (Sankar and Srinivasan, 2017).

IoT Smart Environment

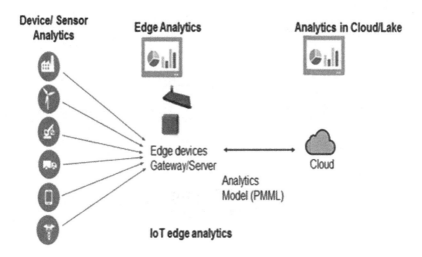

FIGURE 2.5 IoT edge computing analytics.

Here the service and security issues are handled at the edges of the entire IoT framework and hardware like local computing devices will provide solutions for that. Hardware like ARM-based microcontrollers, microchips, NXP (Mandal and Ghantasala, 2019), and other specified microcontrollers are used to handle the huge amount of data collected from the network. This technology helps to enable end-to-end system solutions at low-cost cloud computing environments with low latency. This is also used to make IoT devices very smarter without the concept of the cloud for data storage. Figure 2.5 depicts the details of edge computing in the IoT environment.

2.5.2 IoT Security Issues

IoT with WSN technology provides the data transmission property for storing all the data on to the repository. While transmitting common technologies like Bluetooth, Wi-Fi, LoRa, ZigBee (Mandal and Ghantasala, 2019) are used to create repositories at various places. But accessing those data from low latency and high-speed extracting methods connectivity issues will occur due to the distance. So, cryptography is used to provide security over IoT framework data and encryption, and decryption takes place for security concerns (Sankar and Srinivasan, 2017). There are a lot of protocols and cyber security principles used in these places to access the information in IoT effectively. Figure 2.6 denoted the possible challenges of the IoT framework.

2.5.3 Industrial IoT Challenges

IoT used in industries make a lot of challenges as Industry 4.0 because of huge machines and high-volume data generation. Providing space to store a trillion devices' data with high investment is not possible for small-scale industries now. They are seeking solutions to access different platforms for reducing costs. So, they are approaching the service providers for services and making them vendors for

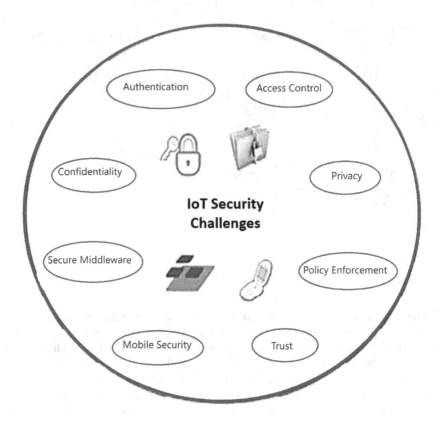

FIGURE 2.6 IoT security issues.

their business. Here security for all the devices and data is the challenge in the IoT environment but recent trends like big data, blockchain, AI, and ML (Shin et al., 2017; Sollins, 2019) are used to solve those challenges with its features. All over the industry trying to solve energy efficiency, business intelligence, data repositories, and automation problems using IoT platforms with service providers available in the IT field. All of them are used cloud and big data technology for data repository purposes but the analytics make them smarter with IoT framework. Figure 2.7 illustrates the challenges faced by industries using IoT.

2.5.4 IoT Cloud Computing Platforms

Cloud computing services namely SaaS, PaaS, and IaaS are used to provide the resource management feature in IoT for scalable input data. Software as a platform in cloud computing has emerged in industries for automation, healthcare, and ERP solutions for business purposes. The issues of software as a service in cloud computing implies vulnerabilities through scalability and heterogeneity over IoT platform. There are other factors also affected by SaaS in IoT like energy consumption, dynamicity, complexity, low latency, and resource utility. So, to improve the software quality in terms of web services will solve the challenges in IoT even though

IoT Smart Environment

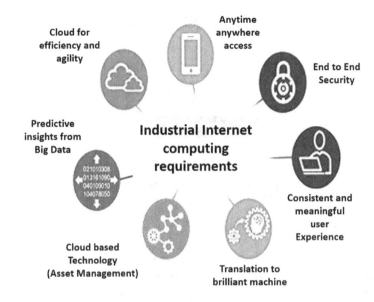

FIGURE 2.7 Industry IoT challenges.

resources are highly utilized in all aspects. Moreover, the IoT framework helps to cloud computing to store the data for analytics, and decision-making by smart sensors or smart devices can be done through these SaaS activities. To protect the data used in remote places devices are taken care of as a web services controller in a centralized place of IoT platform. Figure 2.8 gives the challenges of IoT in cloud computing SaaS platform services.

2.5.5 ML/AI-BASED IoT APPLICATIONS

Data analytics process can do in IoT with ML and AI methods. Deep intelligence is the new approach that helps to connect edge devices and cloud environments to the

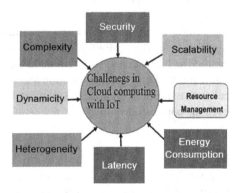

FIGURE 2.8 SaaS challenges faced in IoT.

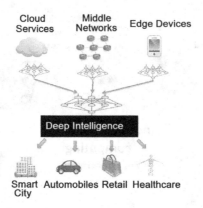

FIGURE 2.9 IoT with ML AI techniques.

smart cities, and healthcare sectors. The private sector is connected to the network by the IoT framework and does its data analytics process using ML concepts. AI is used to select the devices in the network based on their condition throughout the day (Mandal and Ghantasala, 2019; Shin et al., 2017; Sollins, 2019; Karthiprem et al., 2015). Real-time application data analytics processes can avail of these features in IoT to make it efficient. Figure 2.9 shows the deep intelligence method in IoT.

2.5.6 IoT in Sustainability and Climate Change

IoT framework working in various sustainability and climate change projects in and around the world. The projects have increased the usage of recycling products and weather control on earth using the IoT framework. Smart sensors are ruling the world under the IoT framework for the automation and optimization of machines used for real-world applications. The weather control report has generated every second using the IoT framework and futuristic problems like earthquakes, storms, and floods will be identified accurately. Sustainability point of view all the recycled items in the real world have to be centralized and controlled using IoT. Renewable energy and water conservation concepts are improved with the help of IoT smart sensors and are used to monitor (Taha et al., 2018) anywhere from a remote connection. The controlling process of all the devices and machines under IoT became very easy when IoT handled its features. Moreover, the amount of data generation is also very high but handled as big data and stored in a repository for the recycling process. So, the extraction and backup of all information about the entire process is under the control at real-time execution. Figure 2.10 specifies the sustainability and environmental changes in the applications of IoT.

2.5.7 IoT with Blockchain

IoT framework data transmission between the machines and sensors is more secure with the help of encryption and decryption concepts. For providing secure transactions every time the data has to change or be updated from one form to another form

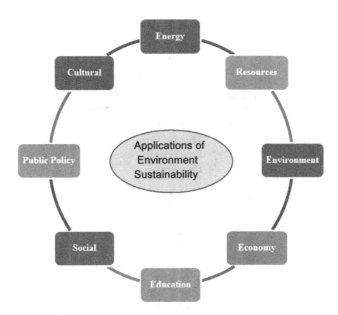

FIGURE 2.10 IoT applications on sustainability and climate change.

during transmission. Due to this if the hacker will give interruption over the network but the data have not been leaked or hacked by the hackers at that time. Blockchain is the new technology used to provide secured transactions over IoT and all the data which are used for processes are encrypted with algorithms. Though the encryption is secured it may cause latency problems during transmission in network-related places. To overcome this blockchains are used as snippet coding transactions by encryption techniques and making every process secure. Figure 2.11 illustrates the significance of the blockchain in IoT.

2.5.8 SERVICES OF CLOUD AND EDGE COMPUTING IN IOT

There are several machines and sensors connected in the IoT framework, and connectivity between the different platforms or networks make conflicts during the carrying of digital data. The platforms create security issues on their edges in the local nodes which are connected internally. So, the nodes are transmitting the data from one place to another place using their two different edges as technologies namely cloud and edge computing (Wen et al., 2018). This will overcome the problem of resource management and energy conservation issues at end-to-end transactions. Here cloud computing stores the data which are generated from all the outside sources whereas edge computing connects the nodes and then the transaction of data have occurred. So, if any hackers are entered any one of the nodes their backup has been taken and stored in their repository moreover it provides encrypted data to hackers (Rana and Kumar, 2019). Figure 2.12 depicts the differences between cloud and edge computing and how it works in the IoT framework.

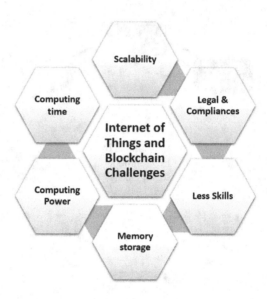

FIGURE 2.11 IoT applications with block chain.

2.5.9 IoT Devices – Sensors Increasingly Becoming Commodity

The devices and sensors connected in the IoT framework have increased rapidly due to real-world applications. But the commodity service for those sensors and devices is increasing day by day for their features. Trillion devices are working continuously and generating data randomly all over the world. The main purpose of the commodity sensor is to implement its capability and caliber of it. So rapid changes and updates in all the sensors used in IoT will improve the efficiency of applications.

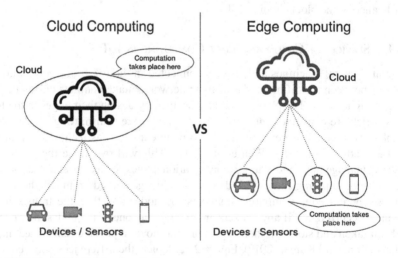

FIGURE 2.12 IoT with cloud and edge computing applications.

IoT Smart Environment 37

2.5.10 LPWAN AND 5G TECHNOLOGIES FOR IoT

Most countries like China, South Korea, Japan, and European countries are tried Low Power Wide Area Networks (LPWAN) for their automation purposes and they have succeeded with good results. It leads to reduce energy consumption and low latency timings for secured transmission at a certain distance. Moreover, 5G is also introduced with high-speed bandwidth and frequency for secured transmission of data over the IoT framework. When discussing these scenarios concerning the time and speed of the data transmission provides good results in real-world applications. But the other factors like the size of data, security, and authentication issues give a poor percentage success rate because it works at a certain distance only. But for IoT framework region in smart city and smart home automation, etc. reveals a good environment for controlling and monitoring the system without a human role at a 75% success rate.

2.6 IoT FUTURE TRENDS

Several technologies and concepts are used in the IoT framework along with their challenges also. But that technological challenges have been solved or rectified by the new approach invented by researchers very frequently. Anyhow, new concepts are ruling the world, especially the digital world, but IoT holds its position strong from the technology begins. It integrates new approaches and algorithms whenever challenges occurred in the current scenario. The analytics part of the entire system in IoT meets difficulties everywhere due to connectivity issues or longer distance data carrying. This will not be solved using all the recent trends so the IoT framework moved on to the next generation of people's expectations. New concepts are coming frequently but their solutions to the existing problems are not up to the levels. The future challenges in IoT resemble the problem of privacy and authentication issues which are occurred during data transmission and other factors like policies, standards, analytics, etc. will also be the challenges in the future. When researching to overcome these issues few concepts are developed using the latest trends and several concepts are changed or updated according to the current scenario. For example, distance will be a key issue in future IoT issues, so the technologies known as LPWAN and 5G will enable this within their range. Automation and optimization are the major problems in IoT and it will be updated in the future with the help of blockchain encryption methods.

2.7 CONCLUSION

The environment of IoT-enabled technologies depicts their difficulties and solutions in data transmission via the network with varied elements. In the IoT framework, creative ways are employed to establish effective communication among devices. Normally, security is the most serious issue with IoT frameworks, resulting in data loss or data corruption when hackers assault the system. When data are sent over long distances via any of the IoT-enabled devices, the IoT framework has connection challenges. At that time, cryptography is utilized to give security to the IoT system environment through authentication, policy, and access standards. This data format has evolved and is now transferred across a greater distance for communication. This method will take security to the next level and bring answers to IoT concerns. Another new technology that will allow the bitcoins technique with a minimum block of data to transit securely is blockchain. The most recent technology, such as ML and AI, are utilized to do data analytics via the cloud platform, allowing computers to make judgments on their own in any circumstance. Deep intelligence is the most recent way to link cloud and edge computing platforms inside the IoT architecture. This connectivity will resolve the challenges that have arisen in the network's local nodes. LPWAN and 5G technologies with a wider variety of frequency patterns are employed for data access and transmission over greater distances. Future IoT frameworks will face several technological obstacles, but the study on current patterns will aid in overcoming these challenges and finding solutions.

REFERENCES

Alcaraz, C. Security and Privacy Trends in the Industrial Internet of Things: Springer, Berlin, Germany, 2019.

Alotaibi, R. A.; Elrefaei, L. A. Improved capacity Arabic text watermarking methods based on open word space. Journal of King Saud University-Computer and Information Sciences. 2018, 30(2), 236–248.

Dabbagh, M.; Rayes, A. Internet of things security and privacy. In Internet of Things from Hype to Reality: Springer, Cham, 2019, 211–238.

Dai, H. N.; Wang, H.; Xu, G.; Wan, J.; Imran, M. Big data analytics for manufacturing internet of things: Opportunities, challenges and enabling technologies. Enterprise Information Systems. 2019, 14(6), 1–25.

Ferrag, M. A.; Derdour, M.; Mukherjee, M.; Derhab, A.; Maglaras, L.; Janicke, H. Blockchain technologies for the internet of things: Research issues and challenges. IEEE Internet of Things Journal. 2018, 6(2), 2188–2204.

Frustaci, M.; Pace, P.; Aloi, G.; Fortino, G. Evaluating critical security issues of the IoT world: Present and future challenges. IEEE Internet of Things Journal. 2017, 5(4), 2483–2495.

Ghafir, I.; Saleem, J.; Hammoudeh, M.; Faour, H.; Prenosil, V.; Jaf, S.; Baker, T. Security threats to critical infrastructure: The human factor. The Journal of Supercomputing. 2018, 74(10), 4986–5002.

Habib, M. A.; Ahmad, M.; Jabbar, S.; Ahmed, S. H.; Rodrigues, J. J. Speeding up the internet of things: Leaiot: A lightweight encryption algorithm toward low-latency communication for the internet of things. IEEE Consumer Electronics Magazine. 2018, 7(6), 31–37.

Hakak, S.; Kamsin, A.; Tayan, O.; Idris, M. Y. I.; Gilkar, G. A. Approaches for preserving content integrity of sensitive online Arabic content: A survey and research challenges. Information Processing & Management. 2019, 6(2), 367–380.

Hakak, S.; Kamsin, A.; Veri, J.; Ritonga, R.; Herawan, T. A framework for authentication of digital Quran. In Information Systems Design and Intelligent Applications: Springer, Singapore, 2018, 752–764.

Jindal, F.; Jamar, R.; Churi, P. Future and challenges of internet of things. International Journal of Computer Science & Information Technology (IJCSIT). 2018, 10, 13–25.

Kamaruddin, N. S.; Kamsin, A.; Por, L. Y.; Rahman, H. A review of text watermarking: Theory, methods, and applications. IEEE Access. 2018, 6, 8011–8028.

Karthiprem, S.; Selvarajan, S.; Sankar, S. Recognizing the moving vehicle while driving on Indian roads. International Journal of Applied Engineering Research. 2015, 10(20), 41471–41477.

Khadam, U.; Iqbal, M. M.; Alruily, M.; Al Ghamdi, M. A.; Ramzan, M.; Almotiri, S. H. Text data security and privacy in the internet of things: Threats, challenges, and future directions. Journal of Wireless Communications and Mobile Computing. 2020, 2020, Article 7105625, 1–15.

Khadam, U.; Iqbal, M. M.; Azam, M. A.; Khalid, S.; Rho, S.; Chilamkurti, N. Digital watermarking technique for text document protection using data mining analysis. IEEE Access. 2019, 7, 64955–64965.

Kiani, F. A survey on management frameworks and open challenges in IoT. Wireless Communications and Mobile Computing. 2018, Article 9857026. https://doi.org/10.1155/2019/9538016.

Lin, J.; Yu, W.; Zhang, N.; Yang, X.; Zhang, H.; Zhao, W. A survey on internet of things: Architecture, enabling technologies, security and privacy, and applications. IEEE Internet of Things Journal. 2017, 4(5), 1125–1142.

Mandal, K.; Ghantasala, G. P. A complete survey on technological challenges of IoT in security and privacy. International Journal of Recent Technology and Engineering (IJRTE). 2019, 7(6S4), 332–334.

Mishra, B.; Jena, D.; Somula, R.; Sankar, S. Secure key storage and access delegation through cloud storage. International Journal of Knowledge and Systems Science (IJKSS). 2020, 11(4), 45–64.

Rana, R.; Kumar, R. Performance analysis of AODV in presence of malicious node. Acta Electron Malaysia. 2019, 3(1), 01–05.

Samanta, S.; Panda, M.; Ramasubbareddy, S.; Sankar, S.; Burgos, D. Spatial-resolution independent object detection framework for aerial imagery. CMC-Computers Materials and Continua. 2021, 68(2), 1937–1948.

Samanta, S.; Singhar, S. S.; Gandomi, A. H.; Ramasubbareddy, S.; Sankar, S. A WiVi based IoT framework for detection of human trafficking victims kept in hideouts. In International Conference on Internet of Things: Springer, Cham, 2020 September, 96–107.

Sankar, S.; Ramasubbareddy, S.; Chen, F.; Gandomi, A. H. Energy-Efficient Cluster-based Routing Protocol in Internet of Things Using Swarm Intelligence. In *2020 IEEE Symposium Series on Computational Intelligence (SSCI)*. 2020a, December, 219–224.

Sankar, S.; Somula, R.; Kumar, R. L.; Srinivasan, P.; Jayanthi, M. A. Trust-aware routing framework for internet of things. International Journal of Knowledge and Systems Science (IJKSS). 2021, 12(1), 48–59.

Sankar, S.; Srinivasan, P. Internet of things (IoT): A survey on empowering technologies, research opportunities and applications. International Journal of Pharmacy and Technology. 2016, 8(4), 26117–26141.

Sankar, S.; Srinivasan, P. Composite metric based energy efficient routing protocol for internet of things. International Journal of Intelligent Engineering and Systems. 2017, 10(5), 278–286.

Sankar, S.; Srinivasan, P. Internet of things based digital lock system. Journal of Computational and Theoretical Nanoscience. 2018a, 15(9–10), 2758–2763.

Sankar, S.; Srinivasan, P. Multi-layer cluster based energy aware routing protocol for internet of things. Cybernetics and Information Technologies. 2018b, 18(3), 75–92.

Sankar, S.; Srinivasan, P. Energy and load aware routing protocol for internet of things. International Journal of Advances in Applied Sciences (IJAAS. 2018c, 7(3), 255–264.

Sankar, S.; Srinivasan, P. Mobility and energy aware routing protocol for healthcare IoT application. Research Journal of Pharmacy and Technology. 2018d, 11(7), 3139–3144.

Sankar, S.; Srinivasan, P. Fuzzy sets based cluster routing protocol for internet of things. International Journal of Fuzzy System Applications (IJFSA). 2019, 8(3), 70–93.

Sankar, S.; Srinivasan, P. Enhancing the mobility support in internet of things. International Journal of Fuzzy System Applications (IJFSA). 2020, 9(4), 1–20.

Sankar, S.; Srinivasan, P.; Luhach, A. K.; Somula, R.; Chilamkurti, N. Energy-aware grid-based data aggregation scheme in routing protocol for agricultural internet of things. Sustainable Computing: Informatics and Systems. 2020b, 28, 100422.

Sankar, S.; Srinivasan, P.; Ramasubbareddy, S.; Balamurugan, B. Energy-aware multipath routing protocol for internet of things using network coding techniques. International Journal of Grid and Utility Computing. 2020c, 11(6), 838–846.

Sennan, S.; Balasubramaniyam, S.; Luhach, A. K.; Ramasubbareddy, S.; Chilamkurti, N.; Nam, Y. Energy and delay aware data aggregation in routing protocol for internet of things. Sensors. 2019, 19(24), 5486.

Sennan, S.; Ramasubbareddy, S.; Balasubramaniyam, S.; Nayyar, A.; Abouhawwash, M.; Hikal, N. A. T2FL-PSO: Type-2 fuzzy logic-based particle swarm optimization algorithm used to maximize the lifetime of internet of things. IEEE Access. 2021, 9(1), 63966–63979.

Sennan, S.; Ramasubbareddy, S.; Luhach, A. K.; Nayyar, A.; Qureshi, B. CT-RPL: Cluster tree based routing protocol to maximize the lifetime of internet of things. Sensors. 2020, 20(20), 5858.

Sennan, S.; Ramasubbareddy, S.; Nayyar, A.; Nam, Y.; Abouhawwash, M. LOA-RPL: Novel energy-efficient routing protocol for the internet of things using lion optimization algorithm to maximize network lifetime. Computers, Materials & Continua. 2021, 69(1), 351–371.

Sennan, S.; Somula, R.; Luhach, A. K.; Deverajan, G. G.; Alnumay, W.; Jhanjhi, N. Z.; Sharma, P. Energy efficient optimal parent selection based routing protocol for internet of things using firefly optimization algorithm. Transactions on Emerging Telecommunications Technologies. 2020, 32(8), 16 pages.

Shin, D.; Sharma, V.; Kim, J.; Kwon, S.; You, I. Secure and efficient protocol for route optimization in PMIPv6-based smart home IoT networks. IEEE Access. 2017, 5, 11100–11117.

Sollins, K. R. IoT big data security and privacy versus innovation. IEEE Internet of Things Journal. 2019, 6(2), 1628–1635.

Taha, A.; Hammad, A. S.; Selim, M. M. A high capacity algorithm for information hiding in Arabic text. Journal of King Saud University-Computer and Information Sciences. 2018, 32(6), 658–665.

Vinodhini, V.; Sathiyabhama, B.; Sankar, S.; Somula, R. A deep structured model for video captioning. International Journal of Gaming and Computer-Mediated Simulations (IJGCMS). 2020, 12(2), 44–56.

Wen, Q.; Wang, Y.; Li, P. Two zero-watermark methods for XML documents. Journal of Real-Time Image Processing. 2018, 14(1), 183–192.

3 Enabling Technologies for Internet of Things (IoT)-Based Smart Cities

*Velleangiri Vinodhini, B. Sathiyabhama,
S. Sankar, S. Ramasubbareddy,
M. R. Sundarakumar, and B. Balamurugan*

3.1 INTRODUCTION

The concept of computer networks evolved from the convergence of two main fields, namely computers and communications. A computer network is a connection between a pair of machines or hardware devices for transferring data and accessing remote information. It facilitates interactive communication over a group of the computer system (Samanta et al., 2021; Sankar et al., 2021; Sennan et al., 2021a). It has high performance brought significant advancements in computing technology. It had classified computing into different environments like personal, timesharing, distributed, cloud, client-server, and cluster computing environments. In the entire environment, communication was carried out by humans only (Revathi et al., 2021; Sankar et al., 2020a; Sennan et al., 2020b). The remarkable concept coined by Kevin Ashton was the Internet of Things (IoT). It replaced humans with machines. According to Kevin Ashton's statement, people have limited accuracy, time, and attention. It makes things difficult for people to capture from the real world. Introducing computers to gather real-world data will reduce the effort and time of humans (Sennan et al., 2021b; Samanta et al., 2021). This creative idea connected the device in a new way. The definition of IoT is evolved from the convergence of multiple technologies, appliances, and devices. It is a technology, where all electronic devices are embedded with a sensor, programmed into hardware and functions without human interaction (Ande et al., 2020; Mishra et al., 2020). IoT systems are heterogeneous based on various hardware platforms and networks. The "thing" in IoT can be any physical object like a camera, mobile, car, light, etc. The elements of IoT are recognition, sensing, networking, and computing(Sankar et al., 2020b; Sankar and Srinivasan, 2020). It interacts with heterogeneous things. The devices or objects can be of different platforms or networks its application of applications like smart home and business automation, driverless cars, healthcare, wearable, security and defense, supply chain, industrial internet, smart farming, and smart city. It has made people's life easy and smart by maximizing services and minimizing human efforts. Like the TCP/IP architecture in networks, the IoT consists of seven-layered architecture. This chapter explains the enabling technologies associated with the IoT seven-layered architecture.

3.2 OVERVIEW OF IoT ARCHITECTURE

IoT has a flexible layered architecture to connect a heap of heterogeneous systems on the network. The enabling techniques of IoT may vary for different domains. This section briefly describes the seven-layered architecture based on the functionality (see Figure 3.1) (Aman et al., 2020; HaddadPajouh et al., 2020).

The IoT environment has no widely accepted single architecture. Each company designs its architecture based on a requirement. Sensors collect and exchange information across a network using protocols, and store information in a centralized data center. If the data is requested, an appropriate method or application is used to retrieve the data (Sankar and Srinivasan, 2019; Sennan et al., 2019). Consistency and reliability are preserved. The top layer is the business layer. It manages the IoT ecosystem as a whole. It offers a benefits model; a decision-making mechanism that brings the consumer together in this framework. The next layer is the application layer. It's directly connected to the end system. Because IoT supports interoperability, it allows different users to use different systems on the end network. The Internet Engineering Task Force (IETF) developed a range of application layer protocols to support IoT applications (Sankar and Srinivasan, 2018a). Integrated service is delivered across a regional network framework. The next layer is data analysis. It provides users with the ability to easily monitor patterns or trends within the information collected by their application. The guidance given by the analytics ensures that the customer is well skilled with the information required to make a profitable

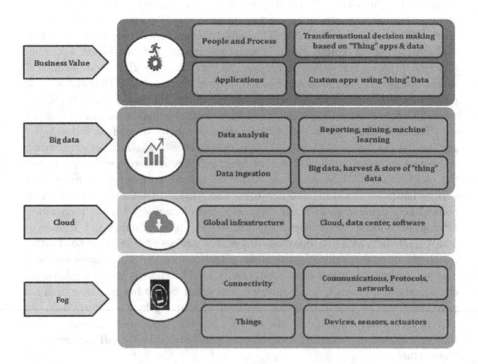

FIGURE 3.1 The seven-layered architecture of IoT.

organization with a clear decision. Data analysis becomes crucial to help the consumer draw important insights without having to do any heavy lifting. Data ingestion needs the collection of events from applications, IoT apps, network and server logs, and transporting them to a data store for processing further (Sankar and Srinivasan, 2018b). The IoT pipeline is created by transferring structured data. If the data is consumed in real-time, it is regulated primarily as soon as the data arrives. As data is processed in different streams, data items are stored in a number of fragments at regular time intervals. Ingestion is the method of moving data to a data processing device. The global infrastructure layer was a recommendation for the global data space in society. It allows systems to interconnect with things based on present and up-and-coming interoperable information and communication technologies (Sankar et al., 2018). The connectivity layer includes a communication protocol and a network where the data can be exchanged. The smart device is continuously gathering data from the user is called the thing layer. When IoT becomes more incorporated into day-to-day life, data analysis becomes crucial to help the consumer draw important insights without having to do any heavy lifting (Washizaki et al., 2020).

3.3 ENABLING TECHNOLOGIES

3.3.1 THINGS LAYER

3.3.1.1 Smart Devices

The physical devices that are embedded with sensors are called things. It may be a smart mobile, laptop, etc. It is responsible for gathering data from the user environment without human interaction are called smart devices. The technology includes radio frequency identification, which can be categorized as passive, active, and semi-passive technology (RFID), The other technology similar to this are quick response (QR code) and infrared readers (IR). Passive applications have minimum consideration (Sankar and Srinivasan, 2018c). They rely on an external source for power supply. RFID is a small chip that enables to track identification. A passive RFID is capable of providing a short-range signal. A semi-passive RFID is powered by batteries which enable to read-write. An Active RFID is a long-range coverage with additional battery life. It is used for real-time applications (Wang et al., 2020).

3.3.1.2 Smart Sensors

A sensor that operates without human intervention is called a smart sensor. The sensor may be integrated into a computer or a standalone sensor, or it may be a wearable sensor. For example, a fitness watch has a sensor that tracks the heart rate (Sankar and Srinivasan, 2018d). These sensors collaborate with all smart devices to fulfill the application requirements (Astill et al., 2020; Sankar and Srinivasan, 2017; Sennan et al. 2020a; Sankar et al., 2020c).

3.3.1.3 Smart Actuators

Physical machines that work automatically on sensed data without human intervention are called smart actuators. It can enable any application to be more effective, fewer employees, and save money. Industry 4.0 is based on smart actuators, where the entire production plant is designed with machines that eliminate human activity.

3.3.2 CONNECTIVITY LAYER

3.3.2.1 Communications

Electronic devices are made smarter by enhancing functionality. IoT has come up with its wireless communication standards. The backbone of IoT is, to couple the applications with network connectivity to share data. The communication protocol is built in to guide data exchange formats, encoding, addressing, and routing the data preserving the sequence control and flow control. It also provides flexibility in the retransmission of lost data. Institute of Electrical and Electronics Engineers (IEEE) derived standard for Wireless Local Area Network (WLAN) 802.11 standard (Mishra et al., 2020).

3.3.2.2 IEEE 802.11Standard-WiFi

Wireless Fidelity (WiFi) is a WLAN. IEEE 802.11 is with different versions such as a, b, d, e, etc. It is a set of WLAN, which supports media access control (MAC) and the lowest layer, the physical layer specification. It offers a minimum range of 20 to 110 m coverage. It uses orthogonal FDM (frequency-division multiplexing). WiFi routers operate on approximately 2.3 GHz and 4 GHz. The 802.11g has 125 feet and 802.11n standards have 235 feet as an indoor range (Karthiprem et al., 2015; Sankar and Srinivasan, 2016).

3.3.2.3 IEEE 802.16 Standards – Wi-Max

The global seamless integration for microwave access (Wi-Max) is static, portable and high-speed internet. It is built on a set of IEEE 802.16 standards. It provides several options in hardware layers (PHY) and MAC. Wireless Metropolitan Area Network (WMAN) connects via unlicensed to a licensed band like 900 MHz, 2.4, and 5.8 GHz, 700 MHz, 2.5 to 3.6 GHz. It is done by orthogonal FDM (frequency-division multiplexing). Table 3.1 represents the range, data rate, and band of the different protocol standards.

TABLE 3.1
IoT Protocol Standards

Protocol	Minimum Range	Power Optimization	Data Rate	Band
Near Field Communication	Personal (<10 m)	Yes	2 Mbps	ISM 2.4 GHz unlicensed
Bluetooth	Contact (<4 cm)	Yes	100 Kbps	ISM 13.56 MHz unlicensed
WiFi	Local(<100 m)	No	>60 Mbps	ISM 2.4 GHz/5 GHz unlicensed
WiMax	Local(<50 km)	No	>70 Mbps	ISM 2.3/2.5/3.5 GHz Licensed
LoRa	Metro(>10 km)	Yes	<50 Kbps	ISM 900 MHz, 868 MHz,433 MHz unlicensed
2G/3G	Metro(>30 km)	No	<2 Mbps	Licensed cellular
4G	Metro(>30 km)	No	>100 Mbps	Licensed cellular
5G	Metro(>30 km)	No	>10 Gbps	Licensed cellular

3.3.2.4 IEEE 802.15.4 Standard – Low Rate – WPAN

IEEE 802.15.4 describes the application of low-rate wireless PAN (Personal Area Networks) called LR-WPANs and manages the IEEE 802.15 task force. This protocol connects millions of devices with less power and storage. These usually work with smart cities and home automation.

3.3.2.5 2G/3G/4G/5G (in Future)

Advances in telecommunications standards have evolved through various generations of cellular area networks. The second generation contains Global System for Mobile Communication (GSM) and Code Division Multiple Access (CDMA). The third generation contains Universal Mobile Telephone System (UMTS) and the 2000 CDMA. These are likely not used by end-users as 4G, 4.5G, 4.75G (including LTE) provide better coverage. The data rate ranges for 4G is from 100Mbs. 5G is expected in the future at large scale.

3.3.2.6 IEEE 802.15.1Standard-Bluetooth

Bluetooth, a wireless device was developed by the IEEE standard 802.15.1. It supports PAN. It provides a very short-range communication of approximately 10 m and provides a relatively lesser bandwidth of approximately 1 to 3 Mbps. Most handheld devices with low power can operate in this network.

3.3.2.7 LoRaWAN R1.0 – Long Range

LoRa (Long Range) is a modulation technique uses spread spectrum with long-lasting battery life operates on 868 and 900 MHz ISM bands. ISM radio bands are globally reserved for commercial, research, and clinical (ISM) use. The data range is approximately 0.2 to 55 kb/s. LoRa is deployed on smart supply chain, smart logistics, etc. It guarantees interoperability between various operators.

3.3.2.8 Light Fidelity – LiFi Technology

Light Fidelity (LiFi) is a technology offering wireless, high-speed data through light communication (VLC). It works with light-emitting diodes (LED) which minimize energy consumption. Its efficiency is limited due to the interference of the light signal. So it is implemented with loss tolerance applications.

3.3.2.9 Software-Defined Network (SDN)

SDN is a flexible, stable, adaptable, and cost-effective networking architecture. SDN-based protected IoT architecture uses protocols for controlling and managing IoT devices, resources, and network entities, like as switches and routers. SDN provides authentication and authorization service to IoT devices (Bera et al., 2017).

3.3.2.10 Blockchain

IoT sends data to private blockchain accounts for the insertion of tamper-proof records on public transactions. IBM blockchain is an example. It enables corporate clients to control and distribute IoT data without any need for central management and monitoring. Every agreement should be checked to prevent conflicts and to ensure that each partner is kept responsible for their position. The first distributed ledger for the exchange of value and data between humans and machines is the IOTA which is the "Internet for everything" (Zheng et al., 2018).

3.3.2.11 New Standards

The ETSI (European Telecommunications Standards Institute) is an autonomous, non-profit standardization body for the communications technology sector that serves European and global business needs. It's based in Sophia-Antipolis, France. ETSI promotes the development and testing of globally applicable technical standards for ICT-enabled systems, software and services that are widely applied in all fields of business and society.

3.3.3 THE GLOBAL INFRASTRUCTURE LAYER

Global infrastructure layer is the third layer. Whenever the node is deployed the centralized service rendered should hike the performance. Considering the performance parameter the deployment is made in cloud and Fog. Whenever the node is too far from the cloud infrastructure, the Fog computation may be considered.

3.3.3.1 Cloud and Fog

IoT applications are usually performed on the cloud infrastructure. The majority of IoT systems are integrated with cloud providers. IoT cloud, a vast range of digital tools, can provide the company with valuable data and insights into clients. Cloud technology is favored for implementations where sensors are used. Sensors that live at various locations are sending data to a centralized cloud for processing. Cloud is versatile because it is an on-demand infrastructure. Cloud providers such as Kaa and DeviceHive support IoT as a service platform (PaaS). IoT service providers use cloud infrastructure as a service (IaaS) to get software as a service (SaaS) from their customers. Fog facilitates to improve the latency while transferring data, improves scalability. If the end-users are too far from the cloud, then Fog can be preferred to extend its service since it is aware of locations and distribution of service. It is capable to connect numerous devices, preserving mobility (Osanaiye et al., 2017).

3.3.3.2 Data Exchange Protocol

To manage an immense amount of data produced by large networks, a technology is needed to efficiently manage the data transmitted to the receiver. IoT has application-layer protocols to do this management. To exchange the data, a node must register as a member of a network; this process is known as subscribing. Once the node has been subscribed publishers may obtain data relating to service; when it is released this is called publish/subscribe process. It is suitable when large nodes are deployed. A request/response process can be carried out for a minimal number of nodes. In this process, the merged node often explicitly demands service-related information and gets a response containing the requested data. When the node enters the system, it assigns network services and also can start exchanging data and function as an IoT system component. Table 3.2 represents IoT data exchange protocol, the platform in which they are deployed and the type of security provided (Sankar and Srinivasan, 2016).

TABLE 3.2
IoT Data Exchange Protocol and IoT Platforms

Protocol	Platform	Integration	Publisher/ Subscriber	Request/ Response	Transport Protocol	Multicast	Security
CoAP	Thing Worx, EVRYTHNG	Rest API	Yes	Yes	UDP	Yes	DTLS
MQTT, MQTT-SN	AWS IoT, IBM Watson, Thing Worx, Bosh IoT Suite, Xively, EVRYTHNG, Kaa	Rest API and Real-time APIs	Yes	No	TCP/IP	Yes	SSL
XMPP	Thing Worx	Rest API	Yes	Yes	TCP	Yes	SSL
AMQP	Bosh IoT Suite	Rest API	Yes	No	TCP	No	SSL
DDS	Thing Worx	Rest API	Yes	No	TCP/UDP	Yes	DTLS

3.3.3.3 Constrained Application Protocol (CoAP)

The IETF designed CoAP, a protocol to work on constrained environments like lower RAM capacity and lossy networks. It is exclusively designed to operate on a wireless personal area network and uses low power (WPAN). It is a web-based protocol, that has an HTTP interface. It allows automation for the machine-to-machine application. It is a client-server model where the client requests and the server responds. It provides good performance even for bulk users. It provides multicast communication where the message is sent to the publisher and the subscriber catches the message.

The CoAP is a reliable and simple protocol, has two layers. The top layer is built by UDP and is called a messaging layer. The other is called the request/response layer. It provides reliable communication by request/response message. Figure 3.2 shows the CoAP message format and Figure 3.3 shows the client-server architecture of CoAP where the interaction is made by REST API. CoAP is transmitted over UDP. It is a fixed-size binary message format. It has options and a payload. The message supports conformable, non-conformable, acknowledgment, and reset. Datagram Transport Layer Security (DTLS) is used as the security protocol. It provides high bandwidth, so the devices connected to it are reliable.

3.3.3.4 Message Queue Telemetry Transport (MQTT)

The MQTT was designed by IBM to reduce the consumption of bandwidth and reliable packet delivery. It reduces the network protocol overhead by minimizing the traffic. MQTT is similar to CoAP by supporting low-power devices, multicast, and publisher/subscriber architecture. It is built on the TCP/IP model. The reliable delivery is considered by three quality of service (QoS) levels such as at least, at most, and exactly. The least level is delivering at least one packet, with no acknowledgment.

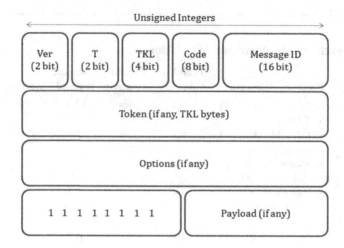

FIGURE 3.2 CoAP message format.

The middle level supports acknowledgment but with duplicate messages. To deliver the message only once and with acknowledgment, few special protocols should be used to reach the highest level of QoS. Figure 3.4 shows the MQTT Broker architecture. The sensor nodes in Figure 3.4 can be a temperature sensor, humidity sensor, etc. Data collected by the sensor node is published to the MQTT Broker rather than the devices directly. The broker will publish the data to the device which has subscribed. Thus a large amount of data can be monitored and controlled by the MQTT Broker.

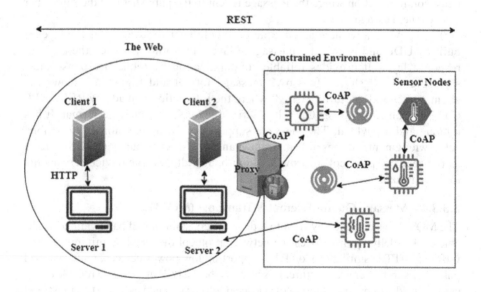

FIGURE 3.3 CoAP architecture.

Enabling Technologies for Internet of Things (IoT)-Based Smart Cities

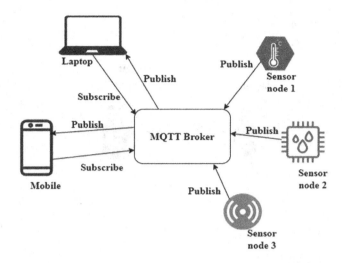

FIGURE 3.4 MQTT broker architecture.

3.3.3.5 Extensible Messaging and Presence Protocol (XMPP)

The most common communication protocol is extensible messaging and presence protocol (XMPP). Its extensible feature allows the protocol to customize based on the functionality. It is a lightweight protocol used in many real-time applications like multi-chat, instant messaging, voice calls, video calls, etc. It uses the transmission control protocol (TCP). It is highly flexible, and scalable with a decentralized architecture. The publisher and subscriber enable event notifications and servers as a foundation for social networking, instant news feeds, and many other real-time interactions. Figure 3.5 represents that XMPP client does not interact directly with another client rather the interaction is done through the XMPP server. Each client is identified by a unique ID called the Jabber ID (JID). It consists of the user, server, and resource. For example, if the user access Facebook, then the "user" in JID is the "Facebook account", the "server" represents the "Facebook server" and the resource represents the "mobile/desktop" on which the application is accessed.

3.3.3.6 Advanced Message Queuing Protocol (AMQP)

AMQP is an open-source, asynchronous messaging standard. It allows secure and interoperable communications between applications and organizations. The protocol is used in client/server communications and the management of IoT apps. It is reliable, compact, secure, and multichannel. The binary protocol offers SASL or TLS authentication and encryption, relying on a transport protocol such as TCP.

The messaging protocol is fast and features assured delivery with confirmation of messages received. It consists of a large and simple component that records and stores communications within a broker operator, along with a collection of policies to connect the components. The AMQP protocol helps the client systems to speak to the dealer and connect with the AMQP model. Figure 3.6 shows the AMQP broker

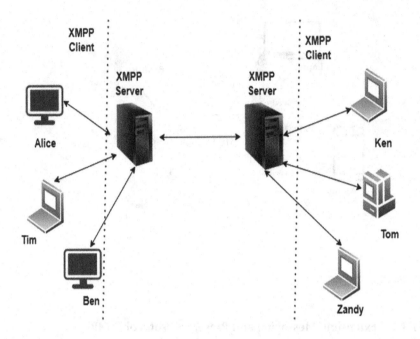

FIGURE 3.5 XMPP client and server.

exchanges the message from the publisher to the subscriber. The queue in the broker model allows guaranteed service to the subscriber (Glaroudis et al., 2020).

3.3.3.7 Data Distribution Service (DDS)

The DDS allows for scalable, real-time, reliable, excessive system results and interoperable statistical change through the submission-subscribe technique. It's less architecture and multicast broker to communicate QoS to high-quality applications.

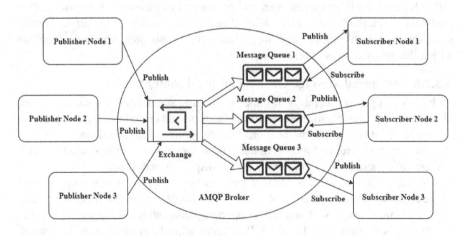

FIGURE 3.6 AMQP broker architecture.

Enabling Technologies for Internet of Things (IoT)-Based Smart Cities 51

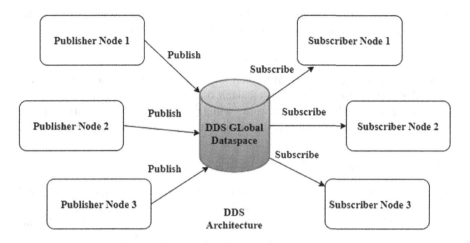

FIGURE 3.7 DDS architecture.

Figure 3.7 shows the publisher to directly communicate with the subscribers in a centralized global data space. DDS is designed to use minimal resources for the system and the network as well.

3.3.4 Data Ingestion Layer

The data ingestion layer is the fourth layer that includes large data, cleaning, streaming, and data management. Apache flume and Apache Nifi are examples of big data ingestion tools. Apache flume provides distributed service. A large amount of data is collected, aggregated, and transferred preserving reliability. A stream of data collected from multiple sources is stored and analyzed in Hadoop. Based on the flow of data streaming, its architecture is made simple. It is robust and tolerant of flaws with many failures and recovery mechanisms. Apache Nifi is an easy-to-use, reliable system, and prevailing for data distribution and processing. It enables tracking data flow, provides seamless service for monitoring and controlling. The data is encrypted by SSL, HTTPS, SSH so it is secure.

3.3.4.1 Storage Tools
3.3.4.1.1 Hadoop Distributed File System
A Java-based file system is used for the secure storage of large data called the Hadoop Distributed File System (HDFS). It is scalable to serve large customers and allows easy access from a large repository. Redundancy is preserved as multiple copies of data are stored on different machines. It is a master/slave architecture model, that allows parallel processing by dividing the data and placing it in different clusters (Nazini and Sasikala, 2019).

3.3.4.1.2 Gluster File System
A Gluster file system (Gluster FS) is an open-source, dependable distributed file system that provides linear scale performance in storage. It can be deployed for

unstructured data. It is scalable and allows distributed storage. Gluster FS can be used in cloud computing, media streaming, and content delivery.

3.3.4.1.3 Amazon Simple Storage Service

Amazon simple storage service (Amazon S3) is a simple, object storage web service. It provides maximum durability and is scalable on widespread. Cloud applications use S3 service. It allows a large data repository with fault recovery. The migration of data is easier than other services. The cost is comparatively reduced when stored on S3.

3.3.5 DATA ANALYSIS

Data analysis is the fifth layer that relates to data investigation, extraction, and machine processing. Most companies use big data analytics to improve productivity, revenue, and lower cost. Apache Hive, Apache Spark SQL, Amazon Redshift, Presto SQL Query Engine are examples of big data query tools (Kamilaris et al., 2017; Samanta et al., 2021; Vinodhini et al., 2020; Karthiprem et al., 2016).

3.3.5.1 Apache Hive

Apache Hive works on Apache Hadoop. It provides a large dataset for analysis, and summarization with an ad-hoc query. The SQL query is used on hive called HiveQL. It is scalable; as data grows many machines are added to enhance the performance.

3.3.5.2 Apache Spark SQL

Apache spark SQL has a spark model for structured data processing. It has a cost-based optimizer. It provides a code generator for fast queries. The scalability is surprisingly high with fault tolerance.

3.3.5.3 Amazon Redshift

Amazon Redshift works on petabytes on cloud. It is used in the processing of data on a query. The database can be created on-demand when an SQL is running. Insight of customers and business is acquired. The overall management is coordinated by redshift. These tasks include the provision of cluster power, monitoring and backup, and the deployment of updates and improvements to the Amazon Redshift system.

3.3.5.4 Presto – SQL Query Engine

Presto is an open-source SQL database engine. It is distributed to run interactive analytical queries of all volumes of a data source. It is designed for commercial use and scales to large extent. It queries data from a relational database. It aims at the analyst who operates the response for each minute. Facebook uses presto against many internal data stores. The employees execute queries greater than 40,000 and scan numerous data. Dropbox and Airbnb are examples of companies using presto.

3.3.6 APPLICATION LAYER

Layer 6 is the base for IoT applications. This layer offers the service needed by the client. The client can use any smart app, such as smartphones, smartwatches, etc. The service rendered by this layer is identity-related service, aggregation service, and collaborative service. The identity-related support offered to the system is constantly monitored and updated. For example, consider the digitization of the medical record. The RFID tag has a patient identity. The data from the tag is gathered and stored in the database. On each visit, the record is updated. The stored information is extracted by web-based or mobile applications. Load delivery between smart grids is an example of data aggregation service. It reviews the information obtained from various sensors. The networks analyze the information and fed the data to the IoT application. Smart home automation requires data to be shared among various smart devices at home. This is an example of a collaborative service. It offers complete control and access at any point in time (Perez et al., 2019).

3.3.7 PEOPLE AND PROCESS LAYER

People and process layers also involve business interaction and decision-making based on IoT computing. Security and data integrity are concerns while implementing IoT on a large scale. Figure 3.8 shows the deployment of security in all technology. Security breaches should be prevented, to reduce the risk of vulnerability. International electrotechnical commission (IEC) is an International Standard. It gives a conformity assessment for all electrical, and electronic technologies. IoT devices should abide by the standards and guidelines of IEC (Ahmed et al., 2018).

3.3.7.1 People

The National Institute of Standard and Technology (NIST) continuously monitors and promotes a new standard to enhance security. A security team is a set of professionals closely working for security. When the vulnerability is detected the security team identifies the impact of severity. It debugs updates and deploys to the customer immediately.

3.3.7.2 Process

There should be a strong security process before deploying a device in real time. The security process should be put into action once a device is discovered with vulnerabilities. In any device deployment, the following Table 3.3 process should be considered to handle the security issues.

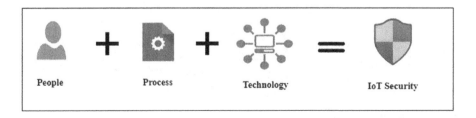

FIGURE 3.8 IoT security.

TABLE 3.3
Security Process

Process	Description
Identifying	Identify the vulnerability
Isolating	Separating it from the network of devices
Validating	Ensuring authenticity
Handling	Handling the vulnerability

3.3.7.3 Technology

Business vendors need trustworthy devices and authentic software. The chain of trust begins at the root and extends to the application layer. Data security plays a vital role. The vendors ensure data security in three forms as shown in Figure 3.9. The securities of data at rest, in use, in motion, are termed as secure storage, secure processing, and secure networking.

Table 3.4 specifies the issues of security. In all IoT infrastructures, various technologies should be implemented to ensure security in all connections. SoC is a facility with a centralized function for an organization. This security team is responsible for security-related issues. The security vendors provide security from the root. To address the layered security model, the hardware isolation defines full SoC in the processor and cache. If hardware has a hypervisor allowing a different operating system, the SoC is on a virtual machine (VM). The software can be isolated in case of vulnerability by the system memory management unit (MMU). The userspace isolation is done on the kernel in the Linux operating system. The encrypted data is stored in the memory in the form of variables and text strings. To protect it, the

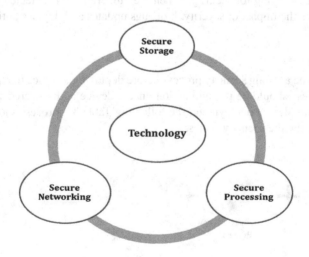

FIGURE 3.9 Data security.

TABLE 3.4
Issues in Data Security

State of data	Issues	Provide security in Hardware Platform	Provide security in Software Platform
Data at rest	Root of trust	SoC (secured operations center)	Yes
	Chain of trust	Boot Loader validated and authenticated	Operating system validated and authenticated
Data in use	Hardware-enforced isolation	Implement multiple SoC	Not applicable
	Software-enforced separation	SoC on embedded hypervisor	Multicore SoC to support virtualization
	Userspace isolation	Not applicable	Memory isolation of data based on applications
	Information or data obfuscation	Configure a firewall, network stress testing, and network penetration analysis	Not applicable
Data in Motion	Device hijack	Not applicable	SoC with crypto engine

developer performs network testing, penetration testing, and configuring a firewall. Data authentication is needed for sending and receiving data. Cryptographic operations should be implemented while routing on the network. The core processor is designed with a crypto engine and supports encryption and decryption. It provides data security at high rates.

3.4 CONCLUSION

The chapter deals with the seven-layered IoT framework which is a modern and widely used technology. The sensors transmit the sensed data through standardized hardware components from the thing layer. To transfer the data a connectivity layer is included on top of the thing layer. The network layer selects the correct protocol for communication and stores in the centralized data store. Information stored may not be structured; the data ingestion layer enables the transfer of unstructured data. Data mining and reporting are performed through a layer of data processing. The user can access the application through a custom mobile application or a web-based application. It is supported by the application layer technologies. Layer 7, the business layer is controlled by the IoT environment that facilitates customers in decision-making. This layer is responsible for the delivery of quality products to the customer. IoT is part of the market with the use of several technologies to facilitate the transformation of industries. Technology continues to evolve due to numerous advances; however, it is crucial to know the features and perceptions of the technology currently available. Since much of the enterprise is driven by IoT, the shortcomings of certain technologies can be addressed in the future.

REFERENCES

Ahmed, S. H.; Rani, S. A hybrid approach, Smart Street use case and future aspects for internet of things in smart cities. Future Generation Computer Systems. 2018, 79, 941–951.

Aman, A. H. M.; Yadegaridehkordi, E.; Attarbashi, Z. S.; Hassan, R.; Park, Y. J. A survey on trend and classification of internet of things reviews. IEEE Access. 2020, 8, 111763–111782.

Ande, R.; Adebisi, B.; Hammoudeh, M.; Saleem, J. Internet of things: Evolution and technologies from a security perspective. Sustainable Cities and Society. 2020, 54, 101728.

Astill, J.; Dara, R. A.; Fraser, E. D.; Roberts, B.; Sharif, S. Smart poultry management: Smart sensors, big data, and the internet of things. Computers and Electronics in Agriculture. 2020, 170, 105291.

Bera, S.; Misra, S.; Vasilakos, A. V. Software-defined networking for internet of things: A survey. IEEE Internet of Things Journal. 2017, 4(6), 1994–2008.

Glaroudis, D.; Iossifides, A.; Chatzimisios, P. Survey, comparison and research challenges of IoT application protocols for smart farming. Computer Networks. 2020, 168, 107037.

HaddadPajouh, H.; Khayami, R.; Dehghantanha, A.; Choo, K. K. R.; Parizi, R. M. AI4SAFE-IoT: An AI-powered secure architecture for edge layer of internet of things. Neural Computing and Applications. 2020, 32(20), 16119–16133.

Kamilaris, A.; Kartakoullis, A.; Prenafeta-Boldú, F. X. A review on the practice of big data analysis in agriculture. Computers and Electronics in Agriculture. 2017, 143, 23–37.

Karthiprem, S.; Selvarajan, S.; Sankar, M. S. Recognizing the moving vehicle while driving on Indian roads. International Journal of Applied Engineering Research. 2016, 10(20), 41471–41477.

Mishra, B.; Jena, D.; Somula, R.; Sankar, S. Secure key storage and access delegation through cloud storage. International Journal of Knowledge and Systems Science (IJKSS). 2020, 11(4), 45–64.

Nazini, H.; Sasikala, T. Simulating aircraft landing and takeoff scheduling in distributed framework environment using Hadoop file system. Cluster Computing. 2019, 22(6), 13463–13471.

Osanaiye, O.; Chen, S.; Yan, Z.; Lu, R.; Choo, K. K. R.; Dlodlo, M. From cloud to fog computing: A review and a conceptual live VM migration framework. IEEE Access. 2017, 5, 8284–8300.

Perez, S.; Hernández-Ramos, J. L.; Raza, S.; Skarmeta, A. Application layer key establishment for end-to-end security in IoT. IEEE Internet of Things Journal. 2019, 7(3), 2117–2128.

Revathi, T. K.; Sathiyabhama, B.; Sankar, S. A deep learning based approach for diagnosing coronary inflammation with multi-scale coronary response dynamic balloon tracking (MSCAR-DBT) based artery segmentation in coronary computed tomography angiography (CCTA). Annals of the Romanian Society for Cell Biology. 2021a, 25(6), 4936–4948.

Samanta, S.; Panda, M.; Ramasubbareddy, S.; Sankar, S.; Burgos, D. Spatial-resolution independent object detection framework for aerial imagery. CMC- Computers Materials and Continua. 2021, 68(2), 1937–1948.

Sankar, S.; Ramasubbareddy, S.; Chen, F.; Gandomi, A. H. Energy-Efficient Cluster-Based Routing Protocol in Internet of Things Using Swarm Intelligence. In 2020 IEEE Symposium Series on Computational Intelligence (SSCI). 2020a, December, 219–224.

Sankar, S.; Somula, R.; Kumar, R. L.; Srinivasan, P.; Jayanthi, M. A. Trust-aware routing framework for internet of things. International Journal of Knowledge and Systems Science (IJKSS). 2021, 12(1), 48–59.

Sankar, S.; Srinivasan, P. Internet of things (IoT): A survey on empowering technologies, research opportunities and applications. International Journal of Pharmacy and Technology. 2016, 8(4), 26117–26141.

Sankar, S.; Srinivasan, P. Composite metric based energy efficient routing protocol for internet of things. International Journal of Intelligent Engineering and Systems. 2017, 10(5), 278–286.

Sankar, S.; Srinivasan, P. Internet of things based digital lock system. Journal of Computational and Theoretical Nanoscience. 2018a, 15(9–10), 2758–2763.

Sankar, S.; Srinivasan, P. Multi-layer cluster based energy aware routing protocol for internet of things. Cybernetics and Information Technologies. 2018b, 18(3), 75–92.

Sankar, S.; Srinivasan, P. Energy and load aware routing protocol for internet of things. International Journal of Advances in Applied Sciences (IJAAS). 2018c, 7(3), 255–264.

Sankar, S.; Srinivasan, P. Mobility and energy aware routing protocol for healthcare IoT application. Research Journal of Pharmacy and Technology. 2018d, 11(7), 3139–3144.

Sankar, S.; Srinivasan, P. Fuzzy sets based cluster routing protocol for internet of things. International Journal of Fuzzy System Applications (IJFSA). 2019, 8(3), 70–93.

Sankar, S.; Srinivasan, P. Enhancing the mobility support in internet of things. International Journal of Fuzzy System Applications (IJFSA). 2020, 9(4), 1–20.

Sankar, S.; Srinivasan, P.; Luhach, A. K.; Somula, R.; Chilamkurti, N. Energy-aware grid-based data aggregation scheme in routing protocol for agricultural internet of things. Sustainable Computing: Informatics and Systems. 2020b, 28, 100422.

Sankar, S.; Srinivasan, P.; Ramasubbareddy, S.; Balamurugan, B. Energy-aware multipath routing protocol for internet of things using network coding techniques. International Journal of Grid and Utility Computing. 2020c, 11(6), 838–846.

Sankar, S.; Srinivasan, P.; Saravanakumar, R. Internet of things based ambient assisted living for elderly people health monitoring. Research Journal of Pharmacy and Technology. 2018, 11(9), 3900–3904.

Sennan, S.; Balasubramaniyam, S.; Luhach, A. K.; Ramasubbareddy, S.; Chilamkurti, N.; Nam, Y. Energy and delay aware data aggregation in routing protocol for internet of things. Sensors. 2019, 19(24), 5486.

Sennan, S.; Ramasubbareddy, S.; Balasubramaniyam, S.; Nayyar, A.; Abouhawwash, M.; Hikal, N. A. T2FL-PSO: Type-2 fuzzy logic-based particle swarm optimization algorithm used to maximize the lifetime of internet of things. IEEE Access. 2021a, 9(1), 63966–63979.

Sennan, S.; Ramasubbareddy, S.; Luhach, A. K.; Nayyar, A.; Qureshi, B. CT-RPL: Cluster tree based routing protocol to maximize the lifetime of internet of things. Sensors. 2020a, 20(20), 5858.

Sennan, S.; Ramasubbareddy, S.; Nayyar, A.; Nam, Y.; Abouhawwash, M. LOA-RPL: Novel energy-efficient routing protocol for the internet of things using lion optimization algorithm to maximize network lifetime. Computers, Materials & Continua. 2021b, 69(1), 351–371.

Sennan, S.; Somula, R.; Luhach, A. K.; Deverajan, G. G.; Alnumay, W.; Jhanjhi, N. Z.; Sharma, P. Energy efficient optimal parent selection based routing protocol for internet of things using firefly optimization algorithm. Transactions on Emerging Telecommunications Technologies. 2020b, 32(8), e4171.

Vinodhini, V.; Sathiyabhama, B.; Sankar, S.; Somula, R. A deep structured model for video captioning. International Journal of Gaming and Computer-Mediated Simulations (IJGCMS). 2020, 12(2), 44–56.

Wang, X. X.; Cao, W. Q.; Cao, M. S.; Yuan, J. Assembling nano–microarchitecture for electromagnetic absorbers and smart devices. Advanced Materials. 2020, 32(36), 2002112.

Washizaki, H.; Ogata, S.; Hazeyama, A.; Okubo, T.; Fernandez, E. B.; Yoshioka, N. Landscape of architecture and design patterns for IoT systems. IEEE Internet of Things Journal. 2020, 7(10), 10091–10101.

Zheng, Z.; Xie, S.; Dai, H. N.; Chen, X.; Wang, H. Blockchain challenges and opportunities: A survey. International Journal of Web and Grid Services. 2018, 14(4), 352–375.

4 Reshape the Sustainable State-of-the-Art Development of Smart Cities

Tarana Afrin Chandel

4.1 INTRODUCTION

During and post-World War II, Dr. Vannevar Bush, US Director of the Office of Scientific Research and Development was serving his duties. He observed that industrial cities of America were facing problems regarding the infrastructure and population to support a 19th- and 20th-century economy [1]. City problems were documented [2–4] and were taken seriously. Decades passed away in resolving these issues. Setting a new framework for financial transformations, and innovations in science and technology, it took decades from redevelopment to usage [5, 6]. A composite (Architecture, Finance) structure along with new innovations in science and technology shaped the model as the concept of smart growth [7] and exploded the transformation toward smart cities [8]. In the 20th century the population also increased rapidly. The population which was 1.5 billion in the 19th century has reached up to 7 billion in 2010 [9]. A better hope of life and efforts to move toward urban civilization, have opened the way for megacities globally. Smart cities appeared as a new idea to overcome the issues regarding the enhanced population, fast development, employment, and enhance lifestyle. All these lightening achievements made people migrate toward cities. In a wide range, smart cities stimulate sustainable financial development. The concept of a smart city varies from people in different cultures. The basic concept of a "smart city" is referred to as the use of digital and ICT-based technology to upgrade the efficiency and performance of urban services and develop new financial opportunities in the cities [10]. With the augmentation of smart cities, initiatives are taken around the world keeping in view about the cost and benefits of smart cities among people, places, and the environment. According to Organization for Economic Cooperation and Development (OECD), smart cities are defined as the initiative or the advancement toward effective digitalization to upgrade the lifestyle of the people and provide sustainable services making a framework of multi-stakeholder process, keeping the environment safe [10]. Smart cities in terms of business are inter-operable systems having 59% sustainability, especially when water and energy are taken into consideration [11] require 57% intercity connection, 46% sustainable transport, 43% private-government collaboration [11]. In this regard, prototype planning was done, turning toward multiple-story towers leading toward

European post-war architecture, which dominated at the end of the 19th century [12]. Implementing this prototype planning in their architecture, forming a new city along with software application, sensor-based embedded system having digital knowledge and information, with improved clean and green environmental surroundings are required to address the constraints of sustainable developments in the urban area [13].

4.1.1 Concept of Smart City Model

The word "smart" refers to the daily usage of products having attractive features such as smart TV (LED, Android) for entertainment, smart cars (e-vehicle, hybrid-vehicle) for transport, and smartphones for connecting the world. At the end of the 19th century, the word smart city became prominent and spread widely toward urban planning [14]. The word "smart city" is generally referred to as innovatively modern cities, efficient for "competitiveness" and "sustainability", by integrating various parameters of development such as economics, mobility, environment, people, living, and governance, becoming self-sustained [14]. At the beginning of 20th century, the 2030 agenda came into existence. The 2030 agenda showed clear development for the bright future of the next generation. Smart cities played a vital role in fulfilling the standards of sustainable development goals (SDGs) [15]. The author Tarana A. Chandel defines sustainable development as 3P, i.e., People, Profit, and Planet [16]. People, Profit, and Planet refer to Social, Economic, and Environment, respectively [16]. United Nations Economic Commission for Europe (UNECE) supported the implementation of the 2030 Agenda in realistic results-oriented in the areas of transportation, ecosystem, finance, renewable energy, commercial, agroforestry, residential, and community [17]. This multidisciplinary framework addressed to integrate the 17 goals of SD and adopt an innovative way of operation that bifurcate the boundaries. It was broadly divided into four major categories given below.

  i. Sustainable use of renewable resources
 ii. Sustainable smart cities for all group of live
iii. Sustainable transportation, smart integration, and connecting globally
 iv. Smart observation and evaluation progress regarding SDGs

4.1.2 Sustainability a Strategic Plan of Smart Cities

The development of smart cities requires the participation of diversified policy makers and stakeholder along with huge resources. This plan is necessary for effective development. The strategic plan aims to analyze a long-term vision that unites the community onto one platform [18] and explain the values of citizen well-being, and economic development [18], paying emphasis on equality, unity, and interaction across various domains [19]. A strategic plan for the development of a smart city has been done considering many perspectives. This strategic plan is divided into three sections, i.e. adoption of smart innovative technology, policy framework and implementation, and governance for a smart city. The deployment of information and communication technologies (ICTs) is very important for developing and improvement of services and infrastructure of urban areas [20, 21]. This technical network improves the infrastructure and enhances the administrative and communication services in the community. Administration and communication services work on the internet of things (IoT), i.e. sensor-based networking, data collection through videos, and internet, from a large and undefined network of people, i.e. feedback of communal services, online complaints, emergency alert systems that send messages to cell phones, smart metering systems, etc. Technical innovations involve engineering, information technology, data science. These fields play a vital role in the research and development of the application-based smart city.

4.1.3 Technology-Based Smart City

Worldwide internet users have increased in the last few decades shown in Figure 4.1. Internet technology along with other communication networking strategies, the IoT became the best platform in developing smart cities. The terminology "Internet of Things" describes an intersection of technologies that permit to access data generated and provided by other devices via wired or wireless internet networks [22].

4.1.3.1 Internet of Things

The major framework for a smart city is based on the IoT. Some have expressed the framework of smart city as three models, four models, and five models. The three models expressed by Balakrishna in 2012 include gathering, sharing, and executing the plan [23]. Some authors have also briefed it as three models, i.e. perception, networking, and application [23, 24]. The four models involve the sensing model, transmission model, processing model, and the application model [25]. Some authors define the same thing using various other nomenclatures. They defined the above

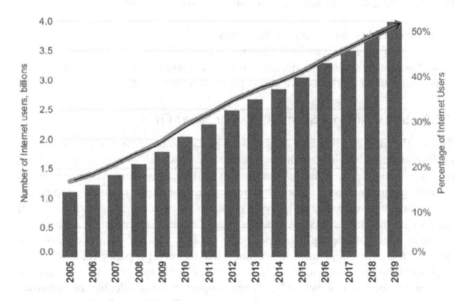

FIGURE 4.1 World wise internet user since 2005.

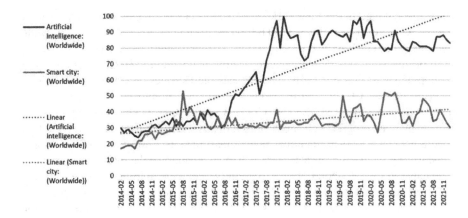

FIGURE 4.2 Smart City and AI increasing from 2014 till 2021 [27].

four models as the data acquisition model, data vitalization model, common data and service model, and application model [26]. Further research suggested it as five model structures, i.e. perception layer, networking layer, servicing layer, utility layer, and trading layer [24].

Figure 4.2 shows increase in utility of IoTs and AI in smart city from 2015 to 2021 [27]. Cities are recognized as smart if they fulfill the criteria of SDGs. Some of the goals outline the smart city.

These are as smart lifestyle, smart learning, smart employment, smart fitness, smart transport, smart ecosystem, smart finance and smart governance as shown in Figure 4.3. In major countries, smart city model is based on the applications of AI.

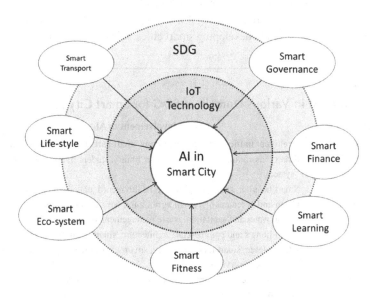

FIGURE 4.3 Sustainable development goals for smart City [27].

The statistical data in the New York Smart Schools Commission Report 2018 shows that the AI company will be of $190 billion by the end of 2025 [27]. The worldwide investment in launching and processing of Artificial Intelligence (AI) system costs $57.6 billion with 75% deployment computing application [27]. China is on top in developing smart cities globally. The city has the facilities of a public place with a metering device, CCTV, and embedded system for monitoring, data mining, all analysis done based on AL [28, 29]. China is having approximately 800 small-scale actively planned programs on smart cities globally. AI in China is working smoothly since 2015 and the government has taken measure steps toward research in AI and transforming AI in China as a new era [30]. AI may upgrade the future of smart cities by enhancing the livelihood of the community people, by joining machine learning (ML), robotics, and many other technologies. AI algorithm is successful depending upon big data for the operation of a useful task. The information is obtained with the help of mechanical and digital technologies that help to store data, process it, and transfer it to give a satisfactory solution to a problem. Table 4.1 shows the involvement of AL in various domains of SDG for the smart city [31, 32].

Approximately more than 1.6 billion IoT components along with other devices were used developing in smart cities in 2017, showing a 39% hike since 2015 [32]. Approximately 3.3 billion IoT tools and components were used [33] with an expansion of 43% [18] in 2018.

4.1.3.2 Artificial Intelligence

AI was introduced by John McCarthy, a computer scientist in 1979 [34]. He defined AL as the combination of science and engineering for designing intelligent machines [35]. It can also be defined as the simulation of human intelligence in a machine, processed to do work according to human action. Brynjolfsson and McAfee have contended in their book "The Race Against the Machine" (Brynjolfsson & Andrew, 2012). Features of AI include ML, nature deep learning (DL), vision, speech, robotics, and expert system. These features are used in developing smart cities. It is mandatory to have big data

TABLE 4.1
Involvement of AI in Various Domain of SDG for Smart City [31, 32]

SDG for Smart City	Involvement of AI
Smart transport	Intelligent traffic management system, smart parking, sustainable transport
Smart learning	Smart classroom, learning tool, smart library, student tracking management system
Smart fitness	Smart hospital, smart healthcare, e-health record, patient monitoring
Smart ecosystem	Clean and green energy, smart agriculture, weather and air quality monitoring, clean water and sanitation, waste management system
Smart live-style	Smart homes and commercial, architecture, smart learning, smart protection
Smart finance	e-Commerce, smart retail shopping, smart supply chain distribution, smart commercial services
Smart governance	e-Governance, policy framework and decision making, disaster management system, urban planning

from sensors applied in IoT for the functioning of smart cities. Big data involves 3V, i.e. volume, velocity, and variety [36]. Volume refers to gigantic data, velocity refers to the speed at which data is processed via algorithm, and variety refers to data collected from different sources. AL and big data are linked together. AL can work on big data collected from different sources as non-human system experiments and imitating human behavior [37]. AL can examine the data and predict it and generate solutions for technologies used in developing smart city. AI in supervised ML, data collected is used to manage AI for finding solutions of raw data. AL will perform the task according to the software program done in it. On the other hand in unsupervised ML, data that is not classified or labeled is sent to AI to find the hidden characteristics in the data [38].

4.1.4 Sustainable Transport for Smart Cities

Responding to mobility in urban areas, transport is very challenging for cities. Corresponding to UNEP, transport is accountable for carbon emission gas globally [39] causing congestion and pollution and declining the standard of life [39]. Sustainable transportation occurs as the basic gear for the development of smart cities, present for long time [40]. Köhler [41] guesses sustainable transport as a global concept. Integration of reliable transport is very must for sustainable development in smart cities. In this regard, connectivity between infrastructure, human, and social capital is very must for better livelihood of the community. Thus we can say that sustainable transport is the driving force for developing smart cities. It leads toward resources for labors, benefiting individuals and globally. Moreover, transport links better lifestyle with community, place, education, and employment and fitness [42–44]. Now we can say that transport impels social and commercial developments. Transport using non-renewable energy sources affects the environment causing pollution in the ecosystem. To overcome these effects, sustainable transport is required. Therefore, sustainable transport shown in the figure is global sustainability that fulfills today's requirements without harming future generations.

The first conference on Global Sustainable Transport held on 26 November 2016 [45]. This conference was opened in Turkmenistan. In this conference, all the stakeholders from the government and private sectors, civil sector, from United Nations and other organizations were brought onto a single universal platform. The conference was carried for two days, discussing about the multiple task of the integrated and crosscutting nature of transport. This meeting was supporting the 2030 Agenda held in Perris for Sustainable development [45]. Such transport can form a bridge in-between people far away, coming closer under one umbrella. Sustainable transport can open the doors toward growth, employment, women empowerment, decline in poverty and death rate, and improve economic development. Sustainable transport is a gateway of smart cities emphasizing safe and healthy ecosystem, resolving many problems, and integrating the policymaker and stakeholders that are associated with the SDGs. Smart cities commonly use ICT to predict and direct urban problems. Smart technology can improve sustainable transport. Latest trendy technologies such as the IoT, AI, and cloud computing (CC) [46] where large data can be stored are bringing new lights toward sustainable transport making a smart city. Addressing the weather change, improving the ecosystem, changing oil-based vehicles to renewable energy-based transport, and using public transport instead of individual vehicles, require big data services. This big data is challenging for public transport operators. The motive of big data is to help industries and academics for identifying the problem and designing a model for sustainable urban transport services. In the current scenario, sustainable transport is dependent on SDGs.. Transport has created jobs, becoming a gateway toward technology application and management, advancing toward the environment safety and healthy ecosystem, improving mobility and connecting people under an umbrella. The latter has enhanced tourism, making better relations globally and improving the economy. These goals helped in achieving equality, respect among people, and multinational benefits along with equalizing the interest among the countries.

Since all countries are not in a situation to invest money in sustainable transport and fulfill the sustainability goals, the speakers involving ministers, upper-class officials, experts and delegates debated for agreement, to reinforce partnerships globally, finding a solution to meet the Paris Agreement on climatic sustainability [45].

4.1.5 Sustainable Smart Learning

Today, it is necessary to understand education as an integral part by which students can achieve technical skills, ability, proficiency, and perspective leading toward an active and powerful process of social marginalization. Smart education is the fourth goal in the 2030 Agenda for SDGs [47], aiming to improve the lifestyle of people via the teaching-learning process [48], solving the major problems of the nation [48]. This goal in turn helps to teach the civilians, how to live in society [49] and bring the nation's economy to top globally. Education also enhances personal and professional development. For the sake of achieving these goals, we require government, private agencies, organizations and entrepreneurs, to take opportunity and responsibility regarding the enforcement and dissemination of sustainability literacy [50, 51]. Considering higher education, which represents cultural heritage, has major liability when incorporating sustainability during curricular design. Such a framework has a huge impact on social, environmental, and economic developments [52–54]. At the same time, the educational move toward sustainability is increasing research in the field of science [55], technology, and management. Twenty-five years earlier, Segovia proposed a pedagogical model known as the "Intelligent Classroom" [56]. An intelligent classroom is based on eight important parameters of the teaching-learning process. These are as follows: goal or motive, methodology, succession, mentor's role, scholar's role, assessment, context, and smart learning environment. In the last parameter, i.e. context we observe a new teaching-learning environment (NTLE) process.

4.1.5.1 Teaching-Learning Environment

NTLE process refers to the maintenance of the environment, equipment capabilities, and culture. For sustainable learning, it is required to safeguard the system at a particular level without draining the resources and preventing from being afraid of system failure. Sustainable education is also referred to as a reliable and continuous technical working system.

The question arises that what additional features make new learning-teaching (NELT) environment. As far as this question is a concern, no particular benchmark is set for it. For basic primary schools, we can use computers, multimedia screens, projectors, and free internet facilities to revolutionize education with IoT [57].

As per the record of National Home Education Research in America, more than 1.6 million children are taking home-school education [57]. IoT-based education systems are transforming education into a new learning-teaching environment. Smart classrooms for schools include advanced technology devices such as whiteboards, interactive learning games, cameras, and technology-based assessments along with feedback options for a sustainable education system [57]. Thus in other words we can say the schools or universities having ICT-based educational infrastructure, physical as well as virtual learning classrooms, self-learning management systems (SLMS) and sustainable resource management systems (SRMS) are considered as smart school/smart-university [58, 59]. A smart teaching-learning environment (STLE) is an intersection of adopting learning content, recommended learning tools or strategies, interaction with users via ubiquitous computing devices and detecting the real-world content as shown in Figure 4.4 [59].

Smart teaching (for teachers) and smart learning environments (for students) conducted by ICT, along with innovative technologies such as IoT and AI, help us achieve quality education, thus fulfilling SDG-4, which is part of the 2030 Agenda, i.e. Parris Agreement on sustainability development goals. This 2030 Agenda has 17 sustainability development goals [58, 60, 61]. The concept of smart learning-teaching having the goal of providing excellent education (goal 4), gender equality (goal 5), clean water and sanitation (goal 6), declining inequality (goal 10), and sustainable societies (goal 11) can be obtained at broad spectrum.

4.1.5.2 Technologies Used for Smart Education

The use of technologies in the education sector is becoming typical globally. IoT, AI, ML, and DL are different technologies often used in the education sector. The application of these technologies is making smart education system.

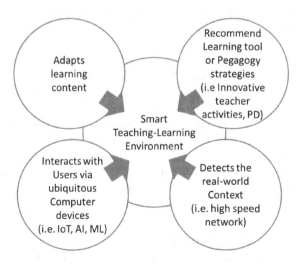

FIGURE 4.4 Green smart teaching-learning environment [59].

4.1.5.2.1 IoT for Sustainable Smart Education

The IoT devices are computing devices which is connected wirelessly to the network, gathering information in the form of data and transmitting it to other devices. IoT consists of both software and hardware. The IoT circuitry is based on sensors as a hardware device and transfers the information data bidirectional online via software technology. IoT benefits students as well as educators. The major benefit of IoT for educators and students is that "one can move beyond the conventional classrooms to the smart classroom environment". Students can learn from videos, webinars, discussion forums, group discussions. IoT helps educator to follow the assignments updated on students' digital planners. It also helps in locating the location of the students and controlling their movements by using real-time cameras making a safe environment and enhancing on-campus security. IoT also helps guardian to track their children's attendance record available in real-time. Wireless communication technology (WCT) and AI are the essential facilitator technologies for the IoT. IoT-based student management system can be done using Bluetooth technology along with IP-based closed-circuit television (CCTV). Entry and exit of students/teachers can be monitored by biometrics, i.e. fingerprinting and face recognition are used to identify the students/teachers. Nowadays, learning can be done in physical mode, online mode, and even hybrid mode. A classroom can be real and virtual. Mobile learning, e-learning, online learning, and distance learning are different pedagogies that have the same concept of making learning easier.

4.1.5.2.2 Machine Learning (ML) for Sustainable Smart Education

ML is a technique used for data analysis. An Australian named GovHack included many projects in the education of which one is ML [62]. This can be used by educators as well as policymakers to identify the risk of students being drop out of the campus. Using mathematical models AI can easily predict outcomes based on previous inputs. With the help of an algorithm, ML permits the system to find out invisible insight without being programmed [62, 63]. The process of learning begins with the data to be observed and comes out with better decisions or results. The process of ML is using an algorithm, can receive data or information, analyze it and finally gives the output within an adequate range. The working process of how ML finds a solution and gives results is shown in Figure 4.5 [63].

ML is classified into supervised ML, unsupervised ML, and reinforced ML shown in Figure 4.5 [64]. In supervised ML, future prediction is done with known input and output data. The result of supervised ML is not continuous, graphically non-linear. Unsupervised ML explores the information whether data is labeled or unlabeled. In reinforced ML, the interaction of information within the environment takes place, giving the result finally. Reinforced ML is used in robots, games, and in determining the position of a ship, i.e. navigation [64]. ML is a sub-array of artificial intelligence (AI) [64]. ML is a subfield of AI as shown in Figure 4.6.

4.1.5.3 Impact of COVID-19 on Education

The coronavirus disease originated in China in December 2019 [65]. The first instance of the virus was found in Wuhan, a city in China. After that, it spread all

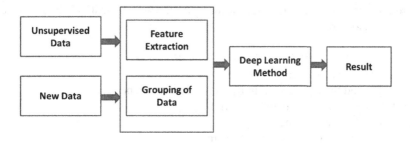

FIGURE 4.5 Process of machine learning [63].

over China. It has affected people irrespective of color, race, and gender [65]. 12 March 2020 was the day when World Health Organization (WHO) announces the coronavirus disease as COVID-19 [66]. WHO also declared a public health emergency of international concern (PHEIC) on 30 March 2020 [66]. Till then the disease spread as current and affected globally. Governments declared the closure of schools, colleges, and universities whether private or government both to maintain social distancing and decline the spreading of the disease. Disease spreading rate was high among children rather than adults, so educational institutes were closed as a preventive action globally. The conventional teaching method/offline mode of teaching, i.e. face-to-face interaction teaching switched toward online teaching. The government, as well as private educational institutes, is following e-learning. Some e-learning platforms are Google classroom, Zoom, WebEx and Skype. Online learning will help teachers as well as students. The evolution of digital systems has a great impact on social and economic development. With the enhancement of innovative technologies and digital devices conventional education system during COVID-19 has turned into digitize education system. Online learning has created hindrances because of internet unavailability, lack of appliances, i.e. mobile and laptops and also the environment of their homes. This has affected the quality of learning, students' understanding and ability and also teachers' performance.

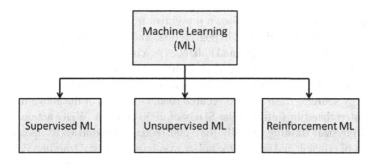

Types of Machine Learning

FIGURE 4.6 Types of machine learning [64].

There are some positive and negative of COVID-19 on education. Some of them are listed below [67].

4.1.5.3.1 Positive Impact of COVID-19 on Education [67]
i. Learning has become personalized and effective
ii. Enhanced the education quality from a future perspective
iii. It is better than conventional learning
iv. Saving time and money
v. No geophysical limitation to learning
vi. Conveyance free education
vii. Application of Learning Management Systems enhanced in education institute
viii. Progress in group debates and teamwork
ix. Request for MOOC and distance learning

4.1.5.3.2 Negative Impact of COVID-19 on Education [67]
i. Obstructed in the educational activity
ii. Lack of electricity power in rural areas
iii. Lack of study environment at home
iv. Lack of knowledge and technology application to parents
v. Parent's responsibility increased toward educating their wards
vi. Unprepared teachers/students for online education
vii. Students' brains are diverted toward entertainment on mobile and social networking
viii. No write up work is given
ix. Lack of nourishment due to school closure
x. Effective assessment of students are difficult
xi. Delay in fee payment
xii. Accessibility toward a digitized world
xiii. Recession effect on employment rate
xiv. Declination in employment opportunities globally

4.1.6 Smart Fitness

Routine physical activities have shown preventive treatment for non-communicable diseases such as cardiac disease, hypoglycemia, and chest cancer. It also prevents from putting on excessive weight, and enhances psychological level and health. Apart from benefits for health from physical activities, societies are more active in making an extra return on investment (ROI) involving declination in the utility of conventional energy sources, fresh air, and secured road. Outcomes of the above-mentioned ROI are interrelated in fulfilling the SDGs of 2030 Agenda. WHO action plan has promoted physical training globally [68], and requested countries to update a new framework of efficient and achievable policy to enhance physical training. Policies for upgrading walking, running, cycling, games, and sport contribute an important role in achieving SDGs of 2030 Agenda. Policies framework on physical training has many benefits in social, cultural, and economic developments, contributing to fulfill many goals of SD such as SD Goal 3 (good health and well-being), as well as other goals including SD Goal 2 (ending all forms of malnutrition); SD Goal 4 (quality education); SD Goal 5 (gender equality); SD Goal 8 (decent work and economic growth), SD Goal 9 (industry, innovation, and infrastructure); SD Goal 10 (reduced inequalities); SD Goal 11 (sustainable cities and communities); SD Goal 12 (responsible production and consumption); SD Goal 13 (climate action); SD Goal 15 (life on land); SD Goal 16 (peace, justice, and strong institutions), and SD Goal 17 (partnerships) [68].

The acceptance of ICT inside the medical sector has driven the idea of electronic health, i.e. e-health, leading toward cost saving and increased efficiency. Devices for e-health are smartphones which had opened the door to mobile-health. Health devices connected to IoT have enhanced the advantages of monitoring facilities globally [69]. Governments in local areas are spending the revenue in establishing ICT-based techno-infrastructure for intelligence and advancing social responsibilities regarding health and care in a sustainable environment. Because of it possibilities for smart cities are endless and many companies such as IBM, INTEL is taking leading initiatives in this sector. These companies have discovered the areas where smart cities play a vital role in community security, green energy, education, social welfare, business development, and Medicare [70]. Smart cities based on ICT technologies provide real-time information about weather conditions, congestion in traffic, environment contamination, antigen concentration, and many more. With the proper utility of healthcare information's, community people can utilize Medicare applications and facilities with awareness. The motive of this study is to generate the idea of smart fitness as the result of interaction between health and smart cities. ICT-based technologies have gained charm toward medical science.

4.1.6.1 e-Healthcare

Commitment for helping and supporting open biomedical problems is due to IoT-based sensor devices and computers. Such devices are electronic health (e-health) [71]. Gunther Eysenbach in 2001 has defined e-health as given below:

> e-health is an emerging field in the intersection of medical informatics, public health and business, referring to health services and information delivered or enhanced through the Internet and related technologies. In a broader sense, the term characterizes not only a technical development, but also a state of mind, a way of thinking, an attitude, and a commitment to networked, global thinking, to improve health care locally, regionally, and worldwide by using information and communication technology.

e-Health not only specifies electrons-health, but many "e's" combine to portray e-health. Some of the "e's" in e-health are mentioned [71] as easily applicable, efficiency, enhanced quality improvement, empowerment of patients, evidence, education and encouragement, enabling, and equality ethics.

Easily applicable: Removing medical fee. Free medical policies focus to decline financial hindrances while obtaining medical services. Free Medicare policies have acquired fame in the last decades. Everyone can apply for free health care. Medicare policies in terms of financial benefits and services are combined.

Efficiency: Efficiency is an important parameter in e-health. High efficiency can be achieved by reducing the medical cost, useless diagnoses, and regular communication between healthcare centers and patients.

Enhanced quality improvement: With the reduction of medical fees and utility of medical policies have enhanced the quality of healthcare services among the community people.

Empowerment of patient: Patient empowerment is termed from the process where the patient acquires better control in taking decisions and actions related to health. Four basic elements of patient empowerment are given as understanding the role of the patient, patient awareness toward knowledge, and engagement of Medicare service provider, patient abilities to gain benefits, and accepting environment to perform the task.

Education and encouragement: Physicians and surgeons should have a thorough knowledge of the medical subject and should know how to maintain better relations with patients. Strong bonding will encourage the patient to share their personal as well as health and financial problems.

Enabling dual-sided communication and transferring information in a quality manner in between patient and Medicare provider.

Equality should be maintained between rich and poor patients. There shouldn't be any gender biasing in handling the patients.

Ethics should be maintained while providing services to online, informed, and private patients.

4.1.7 Smart Ecosystem

The main focus of smart city is based on the environmental impact of the city. Certainly, the important issue is related with the utility of green and clean energy. Thus, saving energy or an alternative source of energy is very must for transport system. Therefore, energy management is one of the best ways for invest and return. Another approach for smart ecosystem in smart cities is IoT. It is based on sensoring large information data. Some of these sensors are environmental pollution sensors, traffic congestion sensors, audio sensors, humidity and weather forecasting sensors, and video capturing cameras aligned all over the city. These sensors will provide information containing problems in the city. However, these problems are resolved by an impressive management system which is challenging itself. Thus, resolving critical issues at the cost of developing a techno-infrastructure. One more important approach of smart ecosystem in smart city is the ability of community people to take part in regular problems of the city. Participation of community people in digital mechanism, sensoring the real-time activities and associative fund management are some of the potentials deployed to make better city management. Now we can say that the city is the data-controlled service provider for social community people along with government and private companies and organizations, data-driven smart ecosystem for smart city [71] is shown in Figure 4.7.

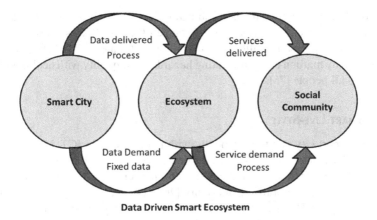

FIGURE 4.7 Data-driven smart ecosystem [71].

These services can be provided with the help of a subcontractor or by the organization making the ecosystem as the foundation with the aid of digital value-added services. This value-added digital system is added by ecosystem managing companies to achieve a different task, providing new processing information. This is how the ecosystem managing companies provide services to the community. The community people or we can say consumers also request the managing companies for adding new salient features for making services easy for smart city. These services are analyzed by some standards taken into account. An ISO standard for smart city is 37120:2014 [72] endorsed by the World Council on City Data [71], consisting of parameter considering the smart city. As big data is generated by the sensors of IoT, the ecosystem administrator further reuses these data for providing further advanced services. For new techno-services, new business models are searched. Some of the business model charge for their services. The reuse of public information has a great impact on the economy of our society. GDP which was 0.25% has increased up to 1.7% [73, 74] while fulfilling some of the market conditions. Undoubtedly, some significant technology will be adapted for smart cities to make better ecosystems.

Thus we can categorize a smart city based on knowledge and designing of a smart ecosystem, planning to implement day-to-day activities easily, and sharing the latest data for further innovations with the help and efforts of our society or community people along with the investments of shareholders. Factors influencing the sustainable and friendly ecosystem are quality of air, contamination in the environment, and improved sanitary system. A healthy ecosystem is categorized by the air quality index (AQI). AQI less than 50 is shown by green color; indicating good air quality with no risk to all people [75]. AQI in the range of 51 to 100 is shown by yellow color; indicating moderate air quality with risk for sensitive persons [75]. AQI in the range of 101 to 150 is shown by orange color, indicating health problems to a sensitive group of people and less effective to the general public [75]. AQI in the range of 151 to 200 is shown by red color; indicating unhealthy air quality and can cause serious health problems [75]. AQI in the range of 201 to 300 is shown by purple color; indicating poor air quality with alarming health alert [75]. AQI greater than 300 is shown by maroon color; indicating hazardous air quality with health problems alarming to all people [75].

4.1.8 SMART LIVE-STYLE

Nowadays, smart city is a trend globally with the progress in ICT. Many countries have passed the proposal for smart city development with the least consideration toward the joyful life of the human. Joy-driven smart city (JDSC) is much better than developing a human-centralized smart city [76]. JDSC requires the investigation of multiple strategies. The strength and weaknesses of society should be on top priority for a smart city as shown in Figure 4.8 [76].

The idea of a smart city is acceptable only with the transformation toward social and economic development [77–79]. Taking initiative toward a smart city, knowledge, and landscape position are top priorities for revitalization for urban financial upgradation and maintaining competencies. However, it is seen that for developing a joyful smart city, minimal attention has been paid on the happiness of the community people. Healthy administrative management and equity will encourage the wisdom of justice and hope to contribute toward the happiness of the community people [80]. A meeting for the happiness and well-being of community people was held in support of UN high level and this report was published by John Helliwell in 2012 [81]. Happiness or well-being is based on emotions, sentiments, and life assessment on particular specifications. However, the basic requirement of the people for sustainability in the city must be available at each hook and corner of the city. Fulfillment of these basic needs enhances joy among everyone. As long as we are living in an enhanced innovative world, a question arises, how advanced

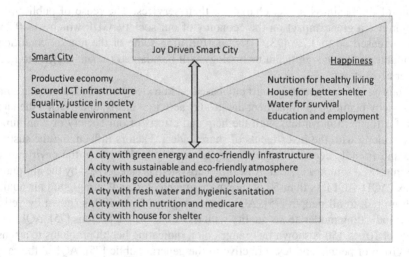

FIGURE 4.8 Joy-driven smart city [76].

innovation can create happiness in society? The motive of this chapter is to outline how we can focus the administrative activities in the city in this techno-world. These activities include tax function, transport sector, municipalities, law and order department, and many more as recorded in OCED better life index (BLI) [82]. The BLI involves income and wealth, household net financial wealth, job and earning, health status, and working and life [82]. Another evaluation for BLI throughout the nation is social progress index (SPI), based on three parameters depending on human need, i.e. house for shelter, better nutrition for improved health and development, water for survival. For human well-being foundation stones are education, health, atmospheric surroundings, communication and innovative technology. For growth and progress in career, a person requires right of freedom, individual rights, and integration. Above-mentioned features are the key source of human happiness and well-being. Factors furnishing toward a happy lifestyle are education, health, and security [76]. Technology awareness directs toward high education, transforming education toward learning, and assisting in achieving their goal [83]. Education in it is a path toward happiness, as it generates employment and enhances the financial status of the people [84]. Safety is also important for happy life. Safety atmosphere for people where offense and safety threats are handled properly, actions are taken immediately with the help of innovative technology corresponding to public safety function [85]. Safety itself is the center of attraction for city and maintaining happy life among each other in a beautiful and smart city [86]. Freedom leads toward a happy life.

4.1.9 Smart Finance

Smart city is categorized by its economy and capability to establish a business. This is the basic reason for the survival of city and people prefer to live as city is a conglomeration of business and finance [87]. In 2007, the economist declared that Homo sapiens are now "Homo urbanus" [88]. Figure 4.9, shows the change in the global

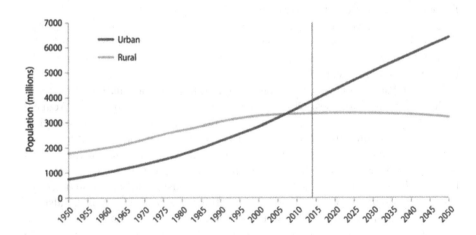

FIGURE 4.9 Change in global population in rural and urban from 1950 to 2050 (Source: United Nations) [89].

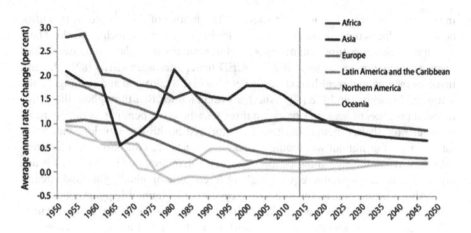

FIGURE 4.10 Transformation toward urbanization region wise from 1950 to 2050 (Source: United Nations) [90].

population of rural and urban from 1950 till 2050 [89]. With the increase in urban population growth, urbanization continued globally, especially with maximum growth performance in the region of Africa and Asia [88].

In 2000, Asia and Africa indicated 37.5% and 34.5% urbanization. In spite of low financial growth in Africa, population growth toward urbanization was nine times from 1950 to 2000 [89]. The statistical data record for the last 65 years shows urbanization in major regions of the world and till 2050 further regions will transform toward urbanization [89]. The percentage transformation toward urbanization region wise is shown in Figure 4.10 [90]. This was the key factor that the city was considered as the motor of the financial growth, fascinating competent and qualified professionals, and national and international investors and contributing toward financial growth nationally and internationally.

The financial layout for the sustainable urbanization evolution is directly associated with the issues of sustainable developments which are associated with 3P, i.e. People, Profit, and Planet [16]. People are related to society, profit to finance, and the planet to the environment [16].

Financial support is very important for entrepreneurs and small- and medium-scale industrial businesses. City enhances the commercial business by frequent interaction with the government and reducing hindrances in developing business. The economic growth in the city is very important for the welfare of the people. For this, the government is working on it and promoting business as its priority. Many cities provide online/internet commerce making it a center for trading platforms. Online trading has made life easier and also enhances the economic condition of the citizen making life happier. Public trust is also an important parameter for the support of finance and public health. Public trust is the biggest evidence for earning and growing economically, leading toward a happy and healthy place to live [81]. A trustworthy atmosphere brings joy and happiness among the citizen. Thus, we can say with pride that trustworthy and corruptionless cities are the measuring parameters for the quality of the nation and the quality of the governance.

These days, there exists a global industrial revolution 4.0 [91]. This industrial revolution 4.0 is the latest paradigm toward industrial exposure and developments. It is based on automation and integrated digital technology and provides bigger information than a physical system [92]. This revolution directs toward enhancing internet potential, i.e. IoT. IoT's interconnect devices provide system automation and large information that can be accessed in real-time [93] as shown in Figure 4.10. The governance for a smart city is directly related to urban development giving rise to industrial revolution 4.0. With fast moving world, the cities are also growing but frequently the system of urban governance is unable to move parallel due to various problem. Some of these problems are high population, building construction compactness, traffic congestion, air pollution, and the environment. The metamorphosis of urban governance is directly proportional to smart city giving rise to new economy [94]. The utility of innovative technology is increased for developing, monitoring, and controlling the activities of the business, traffic, and change in weather conditions [91]. Now the era is known as the era of the digital economy (EDE).

The industrial revolution 4.0 also relates to the trend of increasing internet capacity or the IoT. Based on its terminology, IoT encompasses everything that can be connected from one device to another, with the connection between devices will create system automation, accessing information, analyzing data in real-time, and making actions in response [93]. The incorporation of four industrial technologies has transformed the conventional industry into an adaptable and redesigned smart industry [95]. Smart industry means smart production with innovative technologies such as IoT's AI and data management system. Smart production is related to digital production whose output can be defined as a digital economy. Industries based on ICT, enhance the manufacturing process, making the production cost very low, thus increasing the competition in the market and the growth of the economy. Thus digital marketing (DM) utilizes digital technology and internet facilities in developing business. DM increases the publicity of products and services through real-time data and information of the customer. It also generates sales channels for the products. Information regarding the products such as their cost and product specification is easier and faster, thus increasing the business value. All these are only possible with the digital economy.

4.1.10 Smart Governance

Smart governance is expressed as the examination of complete public areas and duties toward a smart city. Furthermore, it also explores how social and political governance structures can be changed to administer the smart city digitally. Smart governance is two phases of a coin, developing good relations between government-public, reinforcing government-private institutions and integrating all the societies under one umbrella [96]. ICT is now an essential part of our lifestyle. Life is unimaginable without internet facilities and digital technology. Today transport, safety and security, health and fitness, learning and communication are dependent on ICT. The motive of smart governance is to develop a transparent system and strengthen social welfare. The salient feature of smart governance is the utility of ICT tools, the IoT, social media interactions and analysis, private and public cloud, education, employment, and entertainment.

4.1.10.1 Utility of Information and Communication Technology (ICT)

ICT is the base of smart governance. This technology enhances the collaboration level between government departments, government and private sectors and government and society, providing efficient services and sustainability. With the help of ICT, large real-time data and information can be stored giving absolute solutions to the problem. This technology also helps in upbringing the business under the same atmosphere and enhancing the economy of the nation. Good telecom channels such as FM radios, mobile telephones, and satellite communication for receiving and transmitting the information. Transport and travel, video and audio conferencing and security systems require Geographical Information System (GIS) [96].

4.1.10.2 Internet of Things

IoT is based on sensor devices connected to the internet. IoT-based sensory devices are used for monitoring military troops, transportation, drones for defense, temperature control systems, and many more [97]. As written by Bruce Sinclair: IoT is only a transformation toward internet developments. Businesses with IoT technology will lead to economic advancement. IoT-based e-governance ecosystem should guarantee the participation of all policymakers and shareholders as well as the government sector, private sector, technicians, academicians, and consumers.

4.1.10.3 Government and Private Cloud

The birth of the CC model has found a rich and bright base in a smart city in regard to storing large data and information for providing services [98, 99]. CC has gained attention over the past ten years, benefitting where large complex data has to be accessed with restricted resources and economy. The concept of the cloud could act to connect IoT with people on the internet via horizontal and vertical integration [100]. Connecting IoT with CC can be referred to as cloud of things (CoT). Cloud of things is the center for academician and industries globally. It provides services with little investment, awaited progress, high accessibility, excess defect-tolerance ability, endless expandability, and many more.

4.2 CONCLUSION

Developing a smart city with digital technology benefits people globally. It not only enhances the lifestyle but also guarantees sustainability. ICT-based technology provides the solution for big problems saving time, money, and energy. It is also helping in reducing green gas emissions and controlling transportation systems. As the market of smart cities with innovative technology is new for all, it needs a new framework and policymakers to build a new business model and find an easy way to implement sustainability.

REFERENCES

1. Widner, R. R. (1886) Physical renewal of the industrial city. Annals of the American Academy of Political and Social Science 488, 47–57, 1986
2. Brueckner, J. K. (2001) Urban sprawl: Lessons from urban economics. Brookings-Wharton Papers on Urban Affairs 2001(1), 65–97
3. Frey, W. H. (1980) Black in-migration, white flight, and the changing economic base of the Central City. American Journal of Sociology 85(6), 1396–1417
4. Mitra, A. and Mehta, B. (2011) Cities as the engine of growth: Evidence from India. Journal of Urban Planning and Development 137(2), 171–183
5. Wikipedia Contributors (May 21, 2012) "Adaptive reuse," Wikipedia, the free encyclopedia. Wikimedia Foundation, Inc.
6. Kumar, R., Singh, R. C. and Khokher, R. (2022) Framework for modeling, procuring, and building systems for smart city scenarios using blockchain technology and IoT. In The Data-Driven Blockchain Ecosystem (pp 30–50), CRC Press. https://doi.org/10.1201/9781003269281-3
7. Downs, A. (2005) Smart growth: Why we discuss it more than we do it. Journal of the American Planning Association 71(4), 367–378
8. Repko, J. and DeBroux, S. (2012) Smart cities literature review and analysis, MT 598 Spring 2012: Emerging trends in information technology
9. Shukla, P. (2014) Smart cities in India, TERI ENVIS center on renewable energy & environment, The Energy & Resources Institute, Ministry of Urban Development. http://terienvis.nic.in/WriteReadData/links/Smart%20Cities%20in%20India_Report_pagewise-5937837909069130880.pdf
10. Philip, P. (2020) Smart Cities and inclusive growth, OECD, with the support of Ministry of Land Infrastructure and Transport, Korea. https://www.oecd.org/cfe/cities/OECD_Policy_Paper_Smart_Cities_and_Inclusive_Growth.pdf
11. Simpson, P. (2017) Smart cities: Understanding the challenges and opportunities, in association with Philips Lighting survey. https://smartcitiesworld.net/AcuCustom/Sitename/DAM/012/Understanding_the_Challenges_and_Opportunities_of_Smart_Citi.pdf
12. Hall, P. (1988) Cities of Tomorrow: An Intellectual History of Urban Planning and Design Since 1880, Blackwell Publishing: Oxford.
13. Komninos, N. and Mora, L. (2018) Exploring the big picture of smart city research. Scienze Regionali 17, 33–56 © Società editrice il Mulino. https://www.researchgate.net/publication/319598847_Exploring_the_Big_Picture_of_Smart_City_Research
14. Greco, I. and Cresta, A. (2015) A smart planning for smart city: the concept of smart city as an opportunity to re-think the planning models of the contemporary city. Conference: International Conference on Computational Science and Its Applications. https://doi.org/10.1007/978-3-319-21407-8_40
15. Parra-Domínguez, J. et al. (2022) SDGs as one of the drivers of smart City development: The indicator selection process. Smart Cities 5, 1025–1038. MDPI https://doi.org/10.3390/smartcities5030051
16. Chandel, T. A. (2022) Green entrepreneurship and sustainable development, International Perspectives on Value Creation and Sustainability Through Social Entrepreneurship (E-book) (pp. 173–208). https://doi.org/10.4018/978-1-6684-4666-9.ch009
17. Algayerova, O. (2020) People-Smart Sustainable Cities, ECE/INF/2020/3, United Nations Publication, Sales No. E.20.II.E.40, United Nations publication issued by the United Nations Economic Commission for Europe. https://unece.org/sites/default/files/2021-01/SSC%20nexus_web_opt_ENG_0.pdf
18. Kim, S.-C. et al. (2022) Determining strategic priorities for smart City development: Case studies of South Korean and international smart cities. Sustainability 14, 10001. MDPI https://doi.org/10.3390/su141610001

19. Mele, C. (2021) Smart Cities and Sustainability: A Complex and Strategic Issue – The Case of Torino Smart City, IGI Publication (E-book). https://doi.org/10.4018/978-1-7998-7091-3.ch001
20. Hourabi, H. et al. (2012) Understanding smart cities: An integrative framework. 2012 45th Hawaii International Conference on System Sciences, pp. 2289–2297. Maui, HI, USA: IEEE. https://doi.org/10.1109/HICSS.2012.615
21. Frick, K. T. and Kumar, T. (2021) Benchmarking "Smart City" Technology Adoption in California: Developing and Piloting a Data Collection Approach, Publication Date 2021-04-01 Report No.: UC-ITS-2020-31. https://doi.org/10.7922/G2Q23XKF
22. Lin, J. et al. (2017) A survey on Internet of Things: Architecture, enabling technologies, security and privacy, and applications. IEEE Internet of Things Journal 4 (5), 1125–1142. https://doi.org/10.1109/JIOT.2017.2683200
23. Liu, P. and Zhenghong, P. (2014) China's smart city pilots: A progress report. Computer 47(10):72–81. https://www.researchgate.net/publication/270767244_China's_Smart_City_Pilots_A_Progress_Report#fullTextFileContent
24. Gubbi, J. et al. (2013) Internet of Things (IoTs): A vision. Architecture elements, and future direction. Future Generation Computer System 27(7), 1645–1660
25. Silva, B. N. et al. (2018) Inter net of things: A comprehensive review of enabling technologies, architecture, and challenges. IETE Technical Review 35(2), 205–220. https://doi.org/10.1080/02564602.2016.1276416
26. Yin, C. T. et al. (2015) A literature survey on smart cities. Science China Information Sciences 58(10), 1–18. https://doi.org/10.1007/s11432-015-5397-4
27. Herath, H. et al. (2022) Adoption of artificial intelligence in smart cities: A comprehensive review. International Journal of Information Management Data Insights 2(1), 100076
28. Carproti, F. and Liu, D. (2020) Platform urbanism and the Chinese smart city. The co-production and territorialisation of Hangzhou city brain. GeoJournal 87, 1559–1573
29. Dameri, R. P. et al. (2019) Understanding smart city as a glocal strategy: A comparison between. Technology Forecasting and Social Change 142, 26–41
30. Guo (2021) Artificial Intelligence and the mediation of social need in smart city initiative: A critical analysis Presentation, Dublin
31. Bellini, P. et al. (2022) IoT-enable smart cities: A review of concept framework and key technologies, Applied Science 12(3), 1–10
32. Herath, H. et al. (2021) Development of an IoT based system to mitigate the impact of covid-19 pandemic in smart cities. In Machine Intelligence and Data Analytics for Sustainable Future Smart Cities (pp. 287–309), Springer: Cham
33. Khan, M. A. and Salah, K. (2013) IoT security: Review, blockchain solution and open challenges. Future Generation Computer System 82, 395–411
34. Bostrom, N. (2017) Super-Intelligence, Oxford University Press: Oxford
35. Mathur, S. and Modani, U. S. (2016) Smart city a gateway for artificial intelligence in India. 2016 IEEE Student Conference on Electrical, Electronics and Computer Science (SCEECS), pp. 1–3.
36. Graham, N. (2020) Artificial intelligence in smart cities, EMEA, Energy, Government infrastructure Real Estate. https://www.businessgoing.digital/artificial-intelligence-in-smart-cities/
37. Guidance on the AI auditing framework, Draft guidance for consultation, Information Commissioner's Office, 20200214 version 1.0 chrome-extension://efaidnbmnnnib pcajpcglclefindmkaj/https://ico.org.uk/media/2617219/guidance-on-the-ai-auditing-framework-draft-for-consultation.pdf
38. Bamwesigye, D. and Hlavackova, P. (2019) Analysis of sustainable transport for smart cities. Sustainability 11(7), 2140. MDPI. https://www.mdpi.com/2071-1050/11/7/214

39. Mouyal, N. (2021) The future of sustainable transport in cities. E-news website by e-tech. https://etech.iec.ch/issue/2021-02/the-future-of-sustainable-transport-in-cities
40. Transport Must Answer Needs of 'Those Who Have the Least', Says Secretary-General as Global Conference Opens in Turkmenistan. https://www.un.org/press/en/2016/envdev1758.doc.htm (accessed on 26 November 2018)
41. Köhler, J. (2013) Globalization and sustainable development: Case study on international sustainable development. The Journal of Environment and Development 23, 66–100.
42. Rassafi, A. A. and Vaziri, M. (2005) Sustainable transport indicators: Definition and integration. International Journal of Environmental Science and Technology 2, 83–96.
43. Hall, R. P. (June 2002) Introducing the Concept of Sustainable Transportation to the U.S. DOT through the Reauthorization of TEA-21. Master's Thesis, Massachusetts Institute of Technology, Cambridge, MA, USA
44. Hall, R. P. (2006) Understanding and Applying the Concept of Sustainable Development to Transportation Planning and Decision-Making in the US. Ph.D. Thesis, Massachusetts Institute of Technology, Cambridge, MA, USA
45. ASHGABAT (26 November 2016), Global Sustainable Transport Conference, AM & PM Meetings coverage, ENV/DEV/1758. https://press.un.org/en/2016/envdev1758.doc.htm
46. Mulukutla, P. (October 2014) Sustainable mobility for smart cities, WRI India Ross Center, a conference on "Smart Cities" at IIT Madras. https://www.wricitiesindia.org/content/sustainable-mobility-smart-cities
47. UNESCO (2017) Education for Sustainable Development Goals. Learning Objectives; UNESCO: Paris, France. http://unesdoc.unesco.org/images/0024/002474/247444e.pdf (accessed on 20 March 2020)
48. Galán-Casado, D. et al. (28 March 2020) Sustainable environments in education: Results on the effects of the new environments in learning processes of university students. Sustainability 12, 2668. MDPI. https://doi.org/10.3390/su12072668
49. Delors, J. (1996) La educación encierra un tesoro. Informe a la unesco de la Comisión Internacional de el siglo XXI, Santillana/UNESCO: Madrid, Spain
50. Ryan, A. (2011) Education for Sustainable Development and Holistic Curriculum Change: A Review and Guide; HEA: New York, NY, USA. https://sustainability.glos.ac.uk/wp-content/uploads/2017/07/ESD-and-holistic-curriculum-change-EC-16092020.pdf (accessed on 20 March 2020)
51. Winter, J. and Cotton, D. (December 2012) Making the hidden curriculum visible: Sustainability literacy in higher education. Environmental Education Research 18(6), 783–796. https://doi.org/10.1080/13504622.2012.670207
52. Cebrián, G., Pascual, D. and Moraleda, A. (2019) Perception of sustainability competencies amongst Spanish pre-service secondary school teachers. International Journal of Sustainability in Higher Education 20, 1171–1190
53. Cebrián, G. (2018) The I3E model for embedding education for sustainability within higher education Institutions. Environmental Education Research 24, 153–171
54. Rammel, C., Velazquez, L. and Mader, C. (2015) Sustainability assessment in higher education Institutions—What and how?. In Routledge Handbook of Higher Education for Sustainable Development; Barth, M., Michelsen, G., Rieckmann, M., Thomas, I., Eds., (pp. 331–346), Routledge International Handbooks: London, UK
55. Lambrechts, W., Platje, J. and Van Dam, Y. K. (2019) Guest editorial: The university as an arena for sustainability transition. International Journal of Sustainability in Higher Education 20(Special Issue), 1101–1108
56. Segovia, F., Beltrán, J. A. and Martínez, M. R. (1999) El aula inteligente, una experiencia educativa innovadora. Revista Española de Pedagogía 57, 83–109

57. Rajpara, S. (October 9 2020) 6 Internet of Things Benefits in the Education Sector. Updated: May 12, 2021. https://justtotaltech.com/internet-of-things-benefits/
58. Zeeshan, K., Hämäläinen, T. and Neittaanmäki, P. (2022) Internet of things for sustainable smart education: An overview. Sustainability 14, 4293. MDPI. https://doi.org/10.3390/su14074293
59. Freigang, S. (2016) IoT in education by designing smart learning environments, Bosch ConnectWorld Blog. https://blog.bosch-si.com/future-of-work/iot-in-education-by-designing-smart-learning-environments/
60. Américas, E. (2016) Sustainable Development Goals (SDGs): Our history and close relationship. https://www.enelamericas.com/en/investors/a202107-sustainable-development-goals-sdgs-our-history-and-close-relationship.html
61. Thang, S. M., Hall, C., Murugaiah, P. and Azman, H. (2011) Creating and maintaining online communities of practice in Malaysian smart schools: Challenging realities. Educational Action Research 19, 87–105
62. Davison, M. (2012) AI and the classroom: Machine learning in education. https://www.trueinteraction.com/ai-and-the-classroom-machine-learning-in-education/
63. Kuppusamy, P. and Joseph K, S. (November 2021) A deep learning model to smart education system, conference paper, EC-AI ML 2020
64. Dwivedi, D. (May 7, 2018) Machine learning for beginners, Published in Towards Data Science
65. Schleicher, A. (2020) The impact of COVID-19 on education - Insights from Education at a Glance 2020, OECD. https://www.oecd.org/education/the-impact-of-covid-19-on-education-insights-education-at-a-glance-2020.pdf
66. Pareek, T. and Soni, K. (July–December 2020) A comprehensive study on covid-19 pandemic: an impact on school education in India. Amity Journal of Management, VIII(2) (Online).
67. Dar, S. A. and Naseer, A. (2021) Ahmad lone "Impact of COVID 19 on education in India". Kala: The Journal of Indian Art History Congress 26(2(XIV)), 2020–2021
68. Geneva WHO (2018) Global action plan on physical activity 2018–2030: More active people for a healthier world
69. Solanas, A. et al. (August 2014) Smart health: A context-aware health paradigm within smart cities, 0163-6804/14/$25.00 © 2014 IEEE. IEEE Communications Magazine
70. Eysenbach, G. (2021) What is e-health? Journal Medical Internet Research 2001 April–Jun 3(2), e20. https://doi.org/10.2196/jmir.3.2.e20
71. Abellá-Garcíaa, A. et al. (2015) The ecosystem of services around Smart cities: An exploratory analysis. Conference on Enterprise Information Systems/International Conference on Project Management/Conference on Health and Social Care Information Systems and Technologies, CENTERIS/ProjMAN/HCist, October 7–9, 2015. https://doi.org/10.1016/j.procs.2015.08.554
72. **ISO 37120:2014(en),** Sustainable development of communities—Indicators for city services and quality of life, https://www.iso.org/obp/ui/#iso:std:iso:37120:ed-1:v1:en
73. Vickery, G. (2011) Review of recent studies on PSI reuse and related market developments. Information Economics, Paris
74. Dekkers, M., Polman, F., Te Velde, R. and De Vries, M. MEPSIR (2006) Final report of study on exploitation of public sector information – benchmarking of EU framework conditions, Executive summary and Final report. Part 1 and Part 2
75. AirNow, Air Quality Index (AQI) Basics. https://www.airnow.gov/aqi/aqi-basics/#:~:text=Think%20of%20the%20AQI%20as,300%20represents%20hazardous%20air%20quality
76. Zhu, H. et al. (May 2022) How can smart city shape a happier life? The mechanism for developing a happiness driven smart City. Sustainable Cities and Society 80, 103791. https://doi.org/10.1016/j.scs.2022.103791

77. Cheng, C.-Y. et al. (2020) ICT diffusion, financial development, and economic growth: An ross-country analysis. https://doi.org/10.1016/j.econmod.2020.02.008
78. Deloitte (2018) Super Smart City Happier Society with Higher Quality. https://www2.deloitte.com/cn/en/pages/public-sector/articles/super-smart-city.html
79. Vua, K. et al. (March 2020) ICT as a driver of economic growth: A survey of the literature and directions for future research. Telecommunications Policy 44(2), 101922, https://doi.org/10.1016/j.telpol.2020.101922
80. Bin Bishr, A. (2018) Chapter 7, Happy Cities in a Smart World, Global Happiness Policy Report 2018. chrome-extension://efaidnbmnnnibpcajpcglclefindmkaj/https://s3.amazonaws.com/ghc-2018/GHC_Ch7.pdf
81. Helliwell, J. et al. (2017) World Happiness Report 2017. chrome-extension://efaidnbmnnnibpcajpcglclefindmkaj/https://s3.amazonaws.com/happiness-report/2017/HR17.pdf
82. OECD (November 2017) Better Life Index 2017 Definitions and metadata, Sources: OECD calculations based on OECD National Accounts Statistics (database). https://doi.org/10.1787/na-data-en, https://www.oecd.org/statistics/OECD-Better-Life-Index-2017-definitions.pdf
83. Burbules, N. C. et al. (2020) Five trends of education and technology in a sustainable future. Geography and Sustainability 1, 93–97. https://doi.org/10.1016/j.geosus.2020.05.001
84. Powdthavee, N. et al. (2015) What's the good of education on our overall quality of life? A simultaneous equation model of education and life satisfaction for Australia. Journal of Behavioral and Experimental Economics 54, 10–21
85. Hartama et al. (2017) A research framework of disaster traffic management to Smart City Proceedings of the International Conference on Informatics and Computing (ICIC), IEEE, pp. 1–5
86. Cagliero et al. (2015) Monitoring the citizens' perception on urban security in smart city environments. Proceedings of the 31st IEEE International Conference on Data Engineering Workshops, IEEE, pp. 112–116
87. Glaeser, E. L. (2010) Agglomeration Economics, The University of Chicago Press, http://www.nber.org/books/glae08-1, http://www.nber.org/chapters/c7977
88. Vinod Kumar, T. M. and Dahiya, B. (2017) Smart Economy in Smart Cities, Advances in 21st Century Human Settlements, Springer
89. Dahiya, B. (2012) 21st century Asian cities: Unique transformation, unprecedented challenges. Global Asia 7(1), 96–104
90. United Nations, Department of Economic and Social Affairs, Population Division (2014) World Urbanization Prospects: The 2014 Revision, Highlights (ST/ESA/SER.A/352)
91. Tyas, W. P. et al. (2019) Applying smart economy of smart cities in developing world: learnt from Indonesia's home based enterprises. International Conference on Smart City Innovation 2018 IOP Conference Series: Earth and Environmental Science, 248, 012078, IOP Publishing. https://doi.org/10.1088/1755-1315/248/1/012078
92. Schwab, K. (2016) The Fourth Industrial Revolution World Economic Forum®© 2016 – All rights reserved ISBN-13: 978-1-44835-01-9, ISBN-10: 1944835016, REF: 231215, Switzerland. chrome-extension://efaidnbmnnnibpcajpcglclefindmkaj/https://law.unimelb.edu.au/__data/assets/pdf_file/0005/3385454/Schwab-The_Fourth_Industrial_Revolution_Klaus_S.pdf
93. Evans, M. (2018) What is the Internet of Things? WIRED Explains (February 16), https://www.wired.co.uk/article/internet-of-things-what-is-explained-iot
94. Singh, S., Wenzel, G. and Brettschneider, F. (2017) Smart Economy in Smart Cities, 317–322. https://doi.org/10.1007/978-981-10-1610-3
95. Gerekli, İ. et al. (May 2021) Industry 4.0 and smart production. TEM Journal 10(2), 99–805. https://www.temjournal.com/content/102/TEMJournalMay2021_799_805.html

96. Smart City Press (August 21, 2017) Smart Governance for Smart Cities, video & audio conferencing, By smart city
97. Trautman, L. J. et al. (2020) Governance of the Internet of Things (IoT). Jurimetrics, 60(3). https://ssrn.com/abstract=3443973 or http://dx.doi.org/10.2139/ssrn.3443973
98. Petrolo, R. (2012) Towards a smart City based on cloud of things, a survey on the smart City vision and paradigms. Transactions on Emerging Telecommunications Technologies 0000, 00:1–11. https://doi.org/10.1002/ett. https://www.researchgate.net/publication/273389706_Towards_a_smart_city_based_on_cloud_of_things_a_survey_on_the_smart_city_vision_and_paradigms
99. Kakderi, C. et al. (2016) Smart cities and cloud computing: Introduction to the special issue. Journal of Smart Cities 2(1). https://www.researchgate.net/publication/302868995_Smart_Cities_and_Cloud_Computing_Introduction_to_the_special_issue_Journal_of_Smart_Cities
100. Alam, T. (2021) Cloud-based IoT applications and their roles in smart cities. Smart 4, 1196–1219. MDPI. https://doi.org/10.3390/smartcities4030064

5 Emergence of Big Data and Blockchain Technology in Smart City

Meenu Gupta, Rakesh Kumar, Anamika Larhgotra, and Chetanya Ved

5.1 INTRODUCTION

This new world is characterized by innovation through smart-city visions. Smart cities are conceptualized differently in every country. Depending on the country, city, and place, it varies. Every country has its own set of capabilities and requirements, which definitely affects the arrangement as well structure of smart urban dwelling. Since the advent of several cutting-edge technologies in this fast-changing age, innovation has been considered as a prime need to renovate the environment. For instance, during the 1960s when the internet was introduced, it was given an initiation to invent "intelligent" technologies like smartphones, artificial intelligence (AI)-embedded home appliances, smart-mobility, and infrastructural advancements, etc. It reached its zenith in the decade of 1990s, and because of this bitcoin (BTC) was able to be introduced to the world in the end of 20th century [1]. Afterward, BTC gained huge importance and has been responsible for developing and launching blockchain technology in 2008 [2]. Despite this, little literature is available regarding blockchain's uses and challenges in the tourism industry. There have been substantial instances to examine the necessity of using technologies like blockchain technology, big data, and cryptocurrencies in collaboration to develop obsolete settlements. Very profound examples of a smart city setup like Singapore, Amsterdam, and so on are equipped with smart and state-of-the-art appliances for offering services. These include fields like transportation, agriculture, energy-consumption, digital-education, healthcare, telemedicine, sanitation, and governance, and bring smart alterations to suit the dynamism of the modernization. Nowadays, facilitating digitization in everything is the want of technological advancements. Certainly, smart cities will be needed to eliminate space-related issues in the future habitat of citizens worldwide, thus they can be regarded as a great infrastructural asset for them. Also, projects aligned with the approach of smart cities can enable many opportunities for all, especially for cloud infrastructure, IT companies, and smart-device developers. The major challenges posed in large countries like India are the consistent advancement, demographics, and finances. Smart-city concepts, of course, will develop solutions, i.e., environment-friendly and sustainable which further facilitate operational smoothness, efficacy, and better civic amenities for its citizenry [3].

DOI: 10.1201/9781003353034-5

The collaboration of smart city with blockchain technology has unleashed many benefits to the society in terms of economic upliftment, financial safety, infrastructural digitization, data security and many more. Blockchain technology adds more features to the innovative environment brought up by the smart-city approach. Smart cities allow access to services (like transportation, use of appliances, transactions, and delivery/receiving of products) by enabling technology [4]. Embracing technology makes the process of providing services fast enough which in turn facilitates maximum utilization of time for increasing effectiveness of a work. Apart from various modernizations from the smart-city concept, the fear of data-breaches exists which needs rigid structure to pluck all the irregularities in a flexible manner. So for eradicating conundrums to a large extent, enabling blockchain technology along with ongoing services is like a panacea. And there are numerous other fields having huge potential and capacity to break the traditional norm () that covers financial services, education, digitization, healthcare, tourism, insurance, supplies and logistics, security sector, and many more [5].

5.2 LITERATURE SURVEY

There are numerous research papers available on the internet to show the overarching acceptance for the usage of blockchain in the society which is aiding endeavors of developing sectors differently. The insights of different research papers have been used to structure this chapter. Nofer et al. [1] have provided fundamental basis and allied information about blockchain as a ledger system. A briefing has also been provided on the process for carrying out any transaction. Blockchain system features such as security, encryption, decentralization, etc., have been discussed. It has been beneficial in getting the summary about the stated technology. Su et al. [3] have discussed the approach of smart city in a composite manner including numerous activities, focus areas, and allied aspects like challenges, needs, etc. Various things like associated technologies, applicability, credibility, issues, solutions, and so on to create smart cities have been stated. In addition to it the importance of privacy and ways to reduce data violation in a smart city has also been covered. Treiblmaier et al. [4] have shown blockchain as a driving function in development and listed changes brought by blockchain into the cities. This chapter has stated the comprehensive role of the blockchain system in bridging the loopholes encountered in a smart-city approach. It has widely discussed the opportunities unleashed by this technology. Ahmad et al. [6] have described cyber-physical setups as the foundation of the advancements in the traditional-city framework. A list of advantages and challenges associated with transforming traditional cities into smart cities have also been provided. It laid stress on good planning and applications, which could make the smart-city concept an umbrella term. Li [7] has described big data, cloud services, internet of things (IoT), and energy-internet and its respective roles in aiding smart-city concepts. Comparison between the above-stated technologies and solution to the encountered irregularities has been put forward. In addition to its architecture of Peer-to-Peer (p2p) light-heavy backup and pros of facilitating blockchain within smart city has been stated. Also, it briefed about big data and blockchain interconnection and specific about its applications. Hassani et al. [8] have written well-defined briefings about the cryptocurrency, big data, and its utility while in use with blockchain. It was helpful in gaining updates regarding p2p networks and blockchain interconnections. It has shown the role and vitality of digital cryptocurrencies in distributed ledger systems. Hassani et al. [9] have united features of blockchain structure, big data, and digital cryptocurrencies so it can be embedded to the city for making it smarter than before. They cataloged its solitary and composite importance in the computerized economy. It further displayed the interconnection and compatibility within these stated technologies. Nam et al. [10] have discussed the trends and challenges with the traditional society and tourism sector specifically. The author stated that incorporation of blockchain technologies and other modernization aspects has changed the scenarios and brought up many benefits like financially, digitization, security-wise, affordability of accessing services, and many more. Also has specified the usage of blockchain in the tourism sector and in transforming cities. Yu et al. [11] have described a decentralized big data auditing system to improve the reliability and balance of a system without including third-party auditing. It has cited comparison between the decentralized and centralized system as well as showed the analysis in the experimental form. It has also stated that decentralized big data is for increasing the efficacy of data assemblage and the protection provided by the blockchain sequence. Saberi et al. [12] talked about the competence of blockchain in the supply-chain oriented domain and its overarching reach to follow smart-city ideas. Innumerable instances have been shown in which the inclusion of blockchain has rendered efficiency in the supply-chain maintenance and delivered improvement in obtaining desired outcomes quickly. Abbaspour [13] has examined the reach of the blockchain-ledger system in the mining field and enriched knowledge about the same in real-life practices. A systematic illustration of the working of blockchain-ledger and set of states for carrying out a transaction has been described. In addition to that, the author has mentioned the importance of blockchain in other sectors. Kaya Soylu et al. [14] provided information about the various cryptocurrencies existing in the markets like BTC, Ethereum, and Ripple. The effectiveness and vitality of these currencies in the online-sphere have been cited beautifully. The authors have stated the effects and reasons for market volatility associated with cryptocurrencies. They have also defined its utility in the modern era. Some more associated research papers have been discussed in Table 5.1.

5.3 SMART CITY

The prototype of an emerging smart city is concerned with smart urbanization. It is bringing automation to provide services to the residents for enhancing their quality of life. The idea of making a city smart is to renovate the framework of various sectors (includes online banking, transportation, healthcare facilities, energy resources, power distribution, education, emergency warning system, and many more). According to a study, every $1 spent on the technology saves approximately $4 as well as more than $40 trillion is needed world over to transform infrastructural lacunae [6]. The infrastructure of smart-city project considers high-speed internet, interconnectivity, network of sensors and actuators, emerging technologies (like AI, IoT), wireless devices, cloud computing, bringing digitization, and

TABLE 5.1
Associated Research Papers

References	Title	Findings	Observation
Ahmad et al. [6]	Cyber-physical systems and smart cities in India: opportunities, issues, and challenge	• Necessity of cyber-physical systems • Challenges and issues faced by cities from advanced technologies	The author of this paper discussed the importance of advancement of cloud structure and IT field for facilitating smart-city framework. It has listed various challenges, issues, and opportunities that are associated with smart cities and cyber-physical systems.
Eckhoff et al. [5]	Privacy in the smart city—applications, technologies, challenges, and solutions	• Importance of privacy and data security in smart-city structure	The paper has given a comprehensive view about the smart-city project. The various things like associated technologies, applicability, credibility, issues, and solutions, etc., have been provided. The author described the importance of privacy and ways to reduce data violation in a smart city.
Haque et al. [15]	Conceptualizing smart-city applications: requirements, architecture, security issues, and emerging trends.	• About applicability of smart-city concept • Security aspects, privacy concerns and answers to the conundrums • Sustainability • Monitoring devices and scope	This paper has put emphasis on the integrated view and fundamentals to develop a smart city. It laid down the benefits of smart city along with substantial instances that have existed to prove the wonders of automation for the quality of living standards. Also, it talked about incidents related to violation of data security and attached privacy to it. It has mentioned scope of future research, authentication, infrastructural vulnerability, unreasoned interference, and the factor of sustainability.
Alnahari et al. [16]	The application of blockchain technology to smart-city infrastructure	• Implementation of blockchain in different domains for smart-city concept • Effectiveness of blockchain governance	This paper has given a way of utilizing Blockchain sequence to make cities digitally smart, exploring its application in energy, transport, water, construction, and government sectors. Current infrastructure management systems can potentially be upgraded with the help of blockchain.
Balcerzak et al. [17]	Blockchain technology and smart contracts in decentralized governance systems	• Smart contracts in a decentralized system. • Quality assessment tools in smart urbanization.	This paper talked about the decentralized governance systems and combined information articulated on smart contracts and the blockchain ledger. They described PRISMA, i.e., "Preferred Reporting Items for Systematic Reviews and Meta-analyses". AMSTAR, Dedoose, DistillerSR, ROBIS, and SRDR are some of the screening and quality assessment tools that have been analyzed. Also it has shown integration between the ideals and concrete outcomes to create digitized-urban settlement.

automation of household chores. ARUP, a UK services company, has estimated the market value of services installed for bringing automation in the cities. The worth will be $30 billion by 2023 and will further go up to $900 billion a year by 2030 (Pattani and Collins). A cyber-physical system (CPS) is essential as it stores information about the quality of water and air, slot for accommodation, data storage, traffic congestion activities, data delivery, and condition of various infrastructure (i.e., for transportation, energy distribution, healthcare, entertainment, commerce). A wide range of resources are being utilized and deployed every day to give maximum ease in a technological sense. New innovations taking place in the environment have paved the way for huge progress in the IT field and humongous opportunities are provided to its urban dwellers for commerce. A smart-city infrastructure is about "closing the loop" involves recognition, communication, decisive-actions, and monitoring data that is being shared on a daily basis. The two prominent layers of its architect contains "physical" and "cyber" information that with consistent scrutiny takes care of multiple problems related to secrecy, safety, data-storage management, privacy, delivery, installed devices, remedy, etc. The wireless nature of devices makes the process easy and quick.

A digitally smart city is a technologically advanced settlement equipped with digital appliances for carrying out its basic operations. It is an umbrella transformational system able to bring innovation, novelty, dynamism, and a learning factor to the company. For transforming a city and bringing automation, four factors need to be considered. The first factor is the proper installation of electronic and digital appliances and connecting it to the IoT, ICT, and AI for further communication. The second factor is the setting up of relationships between the involved devices and converting the network into a digital environment. The third factor is the utilization of information for allied activities (like for governance, commerce, transaction, etc.) and the last one is to make citizenry friendly with systems which increase the adoption rate of technologies by people. The information and communication technologies (ICT) allow people to improve their thoughts as ICT brings innovation which has the potential to create new ideas [7]. Worldwide, there are numerous examples of cities that have embraced innovation to transform them entirely. Some of these cities are like Chicago, Barcelona, Singapore, Seoul, and Gothenburg. They are highly advanced and have digitized their environment to make urban centers as sustainable centers of growth. They have good water and waste management, traffic facilities, infrastructure, energy efficient systems, e-governance model, and citizen engagement [5]. The concept of a smart city integrates optimization to offer reliable services and to connect citizenry to technology. The greater adoption of ICT and its usage with other technologies like AI, IoT, and blockchain ledger, have improved the delivery of services and available resources are being used to develop more opportunities in the smart-city framework.

The digitized locality can provide remedy for uncountable conundrums. It can alter the drawbacks of rapid urbanization, and be able to reduce pollution, energy consumption, wastage, carbon emissions, infliction of diseases, etc., by using smart devices (like smog towers, repellents, vehicle-pollution check). Smart city is a data-driven plus technology led concept. Smart city needs huge participation of the public to create positive impact and to eliminate irregularities of the traditional world. And the successful implementation of a smart city in one country does not necessarily mean that it can be easily replicated in another city. Globally, the definition of a smart city has different perspectives, for example, the Ministry of Urban Development of the Indian Government is running programs like "Urban Renewal plan" and "The Smart Cities Mission" that are heading toward making India a Smart-nation. India aimed at transforming at least 100 cities and making them technologically advanced [18]. By facilitating smart-city collaboration, accelerating fundable projects with the private sector, securing funding and support from external partners, and facilitating funding and support from ASEAN external partners, the ASEAN Smart Cities Network (ASCN) tries to synergize smart-city development efforts across ASEAN countries [19]. Developing smart cities requires knowledge, creativity, and innovation as a source of urban diplomacy. A report by Merrill Lynch estimates that cryptographic technologies, such as digital currencies and blockchain, will generate approximately 2 trillion Euros for financial institutes by 2025. In order to explore the connection of smart-city management and blockchain, the United Nations has created a network of blockchain cities, a task force that explores how blockchains can be used to run smart cities. According to IBM, Jakarta is transforming itself into a smart city by analyzing an average of 40,000 feedback items per month to make faster decisions to enable smarter IoT services and step toward a smart community. And it emphasizes big data analysis and citizen engagement for such a transformation. In the IDB case study, Singapore's smart-city initiative through the 2014 Smart-Nation-Vision attempted to use ICT, networks and data to build an urban location that is rich in infrastructure and technological progress, better utilization of human resources and smart transportation system, marine equipment, and many more needed to develop a unique smart city [19]. It discusses prioritizing the concept of smart city by streamlining activities related to governance, cost factor, citizenship, national growth, and smart risk taking. So the smart-city perspective is more about the technological advancement that could mitigate the service factor of society and has the potential to be an innovative nation for a better future.

5.4 EMERGING FEATURES OF SMART CITIES

The term smart city implies embedding automation into cites so a variety of services could be accessed through the use of smart appliances and thus gives their residents a much better quality of life. The count of benefits derived from it is huge and the sole basis of the emergence is their feature that has been discussed in the following.

5.4.1 BIG DATA

Big data is analogous to the data with a characteristic capacity, magnitude, or with huge size. It is to illustrate a reservoir or segregation of data that is huge in volume, quantity, specifications, and further enlarging sprightly with growing time span. The growing concern is that existing data management tools (e.g., Oracle, SAP, Microsoft master data services) are not much compatible to deal with such aggregated bulky data with characteristic efficacy [8]. Organizations are making lots of endeavors to maintain data privacy and deal with data-oriented matters. The management of big data is crucial, as unorganized data is properly converted into structured data.

Emergence of Big Data and Blockchain Technology in Smart City

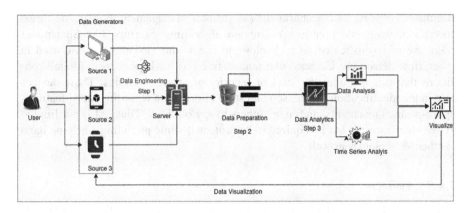

FIGURE 5.1 Big data management.

Thus, any utility can use data smoothly. Figure 5.1 illustrates the process of big data maintenance.

As shown in Figure 5.1, all stages from data generation to data representation work together to maintain big data appropriately. The voluminous data-storage, i.e., big data storage includes everything linked to data and thus formed the term big data. Big data is very essential to analyze how to attract attention that boosts impressive decision-making and prudent dealings in order to enhance growth to the next level. At every step of planning to build a smart city, various embedded fields associated with the same will generate humongous data and without intelligent maintenance of that data will eventually affect the vision of smart city. For this reason big data analytics can help complement the CPS and digital interconnections and thus lead to success in bringing revolution in the technology world. Also by involving big data services, the rank of India in the EODB report has continuously improved and lifted India among the top 50 [18].

5.4.2 Artificial Intelligence

AI is developing a simple, intelligent computerized system by providing instructions to deal with adversities like a human in advance. It is estimated that by 2045 human intelligence will be outclassed by AI and it will bring about 40 million new jobs in the market [20]. AI renders innumerable services to mankind and unlocks fields like quantum technology, deep learning, face recognition, and many more to enhance the predictability of life events. Enabling smart-city solutions, AI brings a range of benefits, including smart energy metering, intelligent traffic-management system, augmenting workforce knowledge and skills, more efficient management of energy, automating routine tasks, smart home appliance, weather detection, and many more. AI adds value to the lifestyles thus automation enhances efficiency and delay in accessing facilities in a smart-city framework is reduced. The improvement in its quality is due to the perfect alignment of NLP, robotics, semantic analysis, ML, and other emerging tech-facilities. Based on big data analytics, ML models can be optimized to derive the predictability of future event occurrence and prepare humans to eliminate the worst case scenario such as disaster management, human life threat prevention, insurance premium estimation algorithms. Startups like Shotspotter make use of sophisticated AI technology to check gun violence and also used in cybercrime detections. Cameras and sensors fitted with AI concepts to foretell conditions that might hinder the level of security of urban dwellings. These devices help in the identification of a person and store information regarding facial features, figure-prints, movements, and many others to spy on them. Thus, AI is not limited to machine learning, but is required to be protected while providing a helping hand to other technologies as well.

5.4.3 FinTech

Apart from the various elements attracting innovation, and automation, finances acts as the backbone which eliminates the major hurdles in the upliftment of a settlement. All decisions and operations are based on finances. A modern and web-based financial system can bring favorable changes in the city, making it a perfect place to live. It focuses on modernizing the banking sector, smart market, public payments, and other finance-related sectors that can support the economy. It has revolutionized by providing services like e-payment, online tax payment, ATM services, etc. According to a study conducted by NSO and FAO 80% delivery of food (like Zomato, Swiggy) or general goods is done through online payment and on average year-on-year increase on online food delivery is 16% to 21% [20]. The same scenario can be seen with taxi services (like uber, ola). As per Indian Statista report India's Internet Penetration rate is almost up by 48% from 2007 to 2021. In forthcoming years it is expected to rise by 75% by 2030, which leads to exponential rise in demand and supply curve for FinTech-based services and providers. It defines the need for strong implementation of FinTech-based architecture [14]. This digitization of transactions was only possible through FinTech. It has the potential to triple its global reach in the coming years. Financial services provided by blockchain technology incorporate asset management, secure record making, quick credit availability, remittances delivery, remuneration to soldiers, minimizing frauds in banks, etc. Property rendering services involve credit availability, card system, smart-device-linked dealing, online delivery, etc. Further in this chapter instances will be cited to show more leverage of blockchain sequence for the FinTech cultures.

5.4.4 E-Governance

By introducing e-governance, the mode of transaction for administering a state has been converted to enable digital culture. It incorporates ICT for carrying out its major operations, i.e., paperless environment and offers sustainability and improves the efficacy of decision-making. Also the idea of smart-city e-governance way initiated in the 1990s and has been continuously supported through finance from World Bank and United Nations [19]. As cities become smarter, traditional ways of governing would not be enough. Smart e-governance is a citizen centric approach embraces more transparency, easy reach, reduces time delays, and many more. The government is adopting new e-governance initiatives (i.e., e-governance with ICT and AI automation) that are linked with the latest technologies and sustainable to the environment. Digital automation makes governance smarter. This would help improve state administration and move toward digital transformation of government. There are four e-governance models in Indian smart cities: government-to-government collaboration (i.e., G2G), government-to-citizenry (i.e., G2C), government-to-employee (G2E), and the last collaboration government-to-business (i.e., G2B). To gather appropriate information above-stated collaborations are necessary for the coordination between the government and the second party. For instance, geospatial web portals like SARTHI, MYGOV in India used to connect citizenry and government for two-way communications and startups like Heat Map, myPolis, and OnSeen, aiding e-governance by aiding recruitment processes, forming climatic-related policies and so on. ("4 Top e-Governance Solutions Impacting Smart Cities") [20].

5.4.5 5G/6G-BASED COMMUNICATION BACKBONE

The growing demand in the networking community is 5G and 6G. 6G internet speed is even faster than 5G which resulted from better performance and increase in bandwidth. Faster connectivity is crucial for smart cities as spontaneous data delivery is necessary to have an efficient ICT structure which is the core ideal of smart city. With higher internet speed such as 40 to 1100 Mbps in 5G or up to 1 Tbps in 6G, tasks will be completed before the usual time span (Holslin). Along with ICT, IoT and ML make the process even more smooth and easy. The 6G era is also expected to beat the cell phone-traffic which is estimated to grow at 608 exabytes per month by the year 2025. And by 2030 this will further rise to more than 5000 exabytes per month for emerging applications [21]. However, a 5G network with AI is needed to create smarter, more efficient, and more secure mobile networks. This is proven (i.e., from a report of qualcomm, Deepsig) as it has created $23 million new jobs, $13Tn global economic output, and 70% increase in investment with AI in network planning [10].

5.4.6 CLOUD COMPUTING

Cloud computing is becoming part of smart-city technologies with its achievements in cost, global scale, speed, performance, efficiency, safety, and reliability. The cloud computing, servers, storage, databases, and numerous application services are easily accessible over the internet with e-payments, making the operations in smart cities connected with the internet to enable fast exchange of information between different users by a blink of an eye. A good internet speed and a consistent flow of data make cloud migration (or cloud storage) easy and reliable. With cloud computing, users can access the system from anywhere on the internet. The objective is to put digital infrastructure on rent and make it affordable for the general public. This helps in remote monitoring and accessing devices from distant places by establishing a network with IoT. Services are available quickly and access is granted for everything connected to ICT. Downtime and maintenance costs are reduced when everything is stored in the cloud; simplifying and speeding up electronic services. The benefits of cloud services will assist in better governing of public data and will take care of confidential matters stored in different repositories. National Information Center, a public sector agency of

Government of India has enabled working on Eucalyptus (i.e., open-source software) for utilizing cloud-based services for disposing government-related operations.

5.4.7 E-Healthcare

In a smart city, healthcare services are digitized by deploying concepts of IoT and AI. e-Healthcare networks thus formed enable telemedicine, telecounseling, e-Prescribing, teledentistry, e-health record, etc., to help humanity dealing with ill health. Having healthy citizens in a smart city means being balanced in every aspect and it makes smart healthcare the most important facility for citizens. Health and smart technology are combined in smart health technology. Nonetheless, according to the market analysis report of Grand View Research, the development in various sectors will comprise 25% of business of smart-city projects by 2025 ("Smart Cities Market Worth $6,965.02 Billion by 2030"). Smart healthcare depends on the robust structure of smart devices. The tools/devices which are digitally operated by doctors for diversified analysis have AI techniques that make the healthcare system as a smart healthcare infrastructure. It gives doctors, healthcare professionals, and researchers an adequate means for examining the disease and the health of the patient [11]. The digital diagnosis forms accurate records and is efficient in saving time, resources, and finances of patients as well as of hospitals. It further displays efficacy in obtaining results from the deployed medications and treatments. Services like telemedicine, virtual consultation, e-health checkups proved to be a panacea during COVID-19 pandemic. The Indian government has a dedicated e-health India portal especially for dealing with e-healthcare services and working for its citizenry. Smart healthcare also makes use of blockchain technology in accurate patient-record-keeping, access control, hospital-related information gathering, hospital-service-maintenance, disease-treatment, insurance processing, and many more [22].

Thus from the above-cited fields of operation, the idea can be formed that the concept of a smart city itself involves so many technologies. Services like smart mobility, cyber security, e-healthcare, online education, IoT-based cloud services, etc., are constantly developing the world and its embedded elements. It is a vast concept with multifold applications. The SUTD (i.e., Singapore University of Technology and Design) and Smart-City Observatory of the IMD-World-Competitiveness-Center got together in 2019 to introduce the very first edition of the IMD-Smart-city-Index. A total of 102 cities have been assessed globally. Singapore, Zurich, and Oslo top the list of the smartest cities of the world as per the 2021 Smart-City index ("IMD Smart-City Index 2019 – IMD Report"). In the upcoming sections, we talk about its integration with blockchain technology as well as security factor involved in making city smarter.

5.5 SECURITY CHALLENGES IN SMART CITY

The idea to have multiple smart cities is to upgrade the structure of cities and utilize emerging technologies in order to enhance productivity and change the lifestyle of the citizenry. But automation and other concepts (like AI, IoT, and ML) are prone to cyber-attacks and put the lives of people in severe risks. By 2025, the demand for IoT devices will be roughly 75 billion, and with the advent of COVID-19 the usage of the internet has seen a huge surge. This has resulted in more illicit cyber activities being done by hackers which have approximately increased to 300 times than usual (reported by FBI, USA). The increasing adoption of IoT poses a greater threat of malicious actors which make systems (that are all attached to the internet) vulnerable. A hacker attack occurs every 39 seconds, and by 2021 it is expected that approximately $6 trillion will be spent on cyber security worldwide [11]. Many of the infamous acts that displace confidential data from the system are phishing, man-in-the-middle attack, ransomware, hacking, distributed denial-of-service (DDoS) attack, and many others. These attacks can cause malfunctioning of power plants, parking garages, EV charging stations, information hijacking, identity theft and other cyber-attacks [23].

The conduct of operations in smart-city framework involves huge data transactions thus availability of smart storage techniques could monitor working adequately. Data is the major element in understanding the system and it helps in coordinating activities of the smart-city project. By safeguarding data, privacy can be prioritized. Thus, maintenance of the source of origination of data captured by various tools (i.e., cameras, sensors, recordings, etc.) carries utmost value. The General Data Protection Rule introduced by the EU concerned with the protection of the personal data of its citizenry thus handling data became easy. Such mandates are also applied by the US government and India is now tightening its data protection laws for internet intermediaries.

The HIPAA Journal has stated estimates regarding data breach (i.e., approximately $4 million is lost globally), and violation of healthcare records (i.e., approximately 10 million records by September 2020 itself). In addition to it, cyber-attacks like social-engineering and phishing attacks have affected more than 60% of the global population (Alder). Also as many as 59% of existing MNCs have encountered botnets, malicious complexities in their system and attacks related to denial of service has gone above 50% ("15 Alarming Cyber Security Facts and Stats | Cybint"). The overlapping facts of security threats are not only limited to technology companies, but have caused danger to key infrastructures such as defense, healthcare, oil markets, banking, and pharmaceutical sector. Increasing dependence on digital setup is exposing various cracks that threaten people's security and open the door to the dark web for illegal activities. The alert systems for emergency warning installed in Baltimore, Dallas, and other Texas cities had been hacked which compromised delivery of information for a few days. This was a major breach of security and was highlighted in the UC Berkeley report. Various examples of cyber-attacks that can be seen around the world are the 2022 cyber-attacks in Ukraine (performed during the prelude to the Russian invasion of Ukrainian territories), the 2019 Pegasus attack, the WHO said in March 2020 that hackers leaked credentials from WHO staff, WannaCry ransomware attack of 2017, and many more. These are just a few cases, but the reality of the reach of a cyber-attack is quite terrifying. To this end, emphasis should be put on the credibility of system software and robust CPS need to be deployed for critical infrastructure protection [24]. Proper modifications in policies will help in coping up with the dynamic environment and will derive new concepts by merging features of IoT and AI to enhance security of the system. Measures taken to secure smart cities include third-party access monitoring, security controls, authentication, encryption-decryption, encryption, application-layer maintenance, and ultra DDoS protection.

5.6 SAFETY CHALLENGES TACKLED BY BLOCKCHAIN OVER BIG DATA

Blockchain is considered as a revolutionary core technology, seems to be a disruptive technology as undoubtedly carries a large potential to destabilize the IT sector in wholesome. Although many researchers are moving ahead to unveil the varied scope of blockchain, the blockchain laden research is yet not able to surpass its genesis stage. The blockchain erection is meddle-proof, kind of distributed ledger representation which writes down various transactions on to the blocks using either public network or private Gnutella network (i.e., a kind of private p2p network) [8]. A cryptographer David Chaum is the first one who put forward protocols regarding blockchain in his 1982 thesis referring ongoing scenarios encountering computer systems and named it as "Computer Systems Established, Maintained, and Trusted by Mutually Suspicious Groups" [1]. In 1991, researchers introduced further information regarding this technology, and then in the year 2008, enigmatic Satoshi Nakamoto, an incognito for either a person or a faction in wholesomeness, published a publication named "Bitcoin: A Peer-to-Peer Electronic Cash System," and in 2009, imagination found its form via establishing the first successful transaction in BTC that occurred within a technologist named "Hal Finney" and an anonymous figure named "Satoshi Nakamoto" [25]. Recorded information secured within the nodes is shared with all linked parties, forming a robust ledger. This composition has a sequentially linked pattern with hash-function marked (using cryptography formulations) blocks, details of profuse asset reciprocities that are carried out within the involved compeers of the grid. Also, blockchain-type representation is a well-formed composition that stows conduction statements or prior interconnections, also known as the block, further stretched in continuous form or representation in databases, known as the "chain," that is, a network of blocks or nodes connected through p2p patterns [21]. Apiece block holds a distinguished encrypted-hash-value (i.e., via cryptography) pointing to the forgoing block (i.e., history), a stamped-marked-record, and interchange details (generally represented as a Merkle tree to depict path to conduct a transaction) [26]. Characteristically, this type of storage sequence forms a "logical ledger" (profoundly known with digital-ledger phrase).

Blockchain technology is a decentralized as well as distributed database of records that is sometimes referred to as a shared ledger system either public or private type ledger for executing different digital interactions among the participating agents in the system of blockchain linkages. Its past details are essential in recalling the entire history of distributed ledger through hash function. Blockchain technology is originally preferred as a variant of design undertaking the operation of the digital currency, i.e., associated with BTC laden businesses. The inventor of BTC has not at all accepted the name "blockchain" within his publication. The blockchain technology is generally composed of six basic building elements (i.e., decentralization, sovereignty, transparency, immutable, and anonymity) [10]. Implementation of blockchain sequencing

Emergence of Big Data and Blockchain Technology in Smart City

is done with the usage of three major technologies (i.e., p2p network, cryptography using private-member-key, and with node-structuring protocols). Approbation of every executed transaction present in the ledger is done by ensuring credibility of the owner's digital signature (i.e., a kind of digital certification), used for authenticating recognition of any transaction in a way to avoid tampering and any malfunction to the data [11]. While appending a new node or record to the blockchain authentication of that data is ensured then the confirmation of its alignment is given, eventually forming a chaining structure in a long way with each node having each-other's copy. In this regard data added to the digital ledger system which follows a decentralized procedure ensures safety and durability of the information added previously. The voluminous data formed while transacting information within the peers and it needs proper maintenance as well as regular tracking to distinguish linkages [27]. The continuous addition of data is organized with the help of big data management. Thus big data and blockchain systems cannot be segregated, is shown in Figure 5.2.

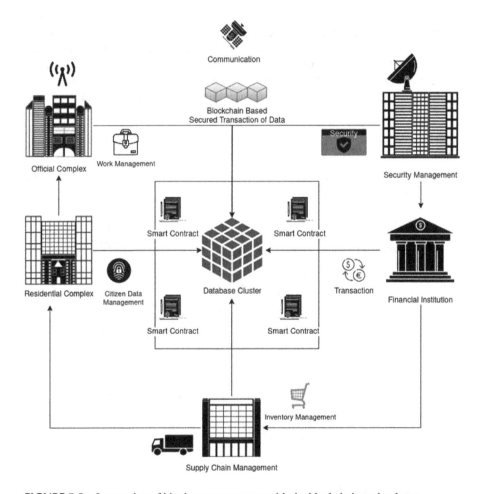

FIGURE 5.2 Integration of big data management with the blockchain technology.

The integration of the two technologies (i.e., big data and blockchain) supplements the supply-chain structure, which in turn segregates the operations of every node of the ledger. Figure 5.2 shows this segregation and depicts the process of block-authentication. Also in the blockchain approach for authenticating the addition of a new transaction, an agent first checks its data thoroughly, then only enters the sequence and thus progresses toward its destination [28]. After that newly processed record is then broadcasted to the blockchain network for further screening processes like accreditation, attestation, validation, auditing, and scrutinizing. If approval for this new transaction is granted by majority nodes as per the determined rules then the transaction would confirm its space in the chain, information about the same is distributed among the nodes as copies. Ad interim, the critical feature of blockchain technology, i.e., lacking the involvement of any mediator, smart contract is able to accept the transactions and performance won't be affected.

The digital ledger system resembles a decentralized setup (like cloud network) as shown in Figure 5.3, which can be accessed by different individuals simultaneously for accessing services and being portable to use from anywhere. The amazing part is that every time when changes are easily encountered by the owner in its database and trace the individual by its email address and other details recorded by the digital platform, only requisite users can make changes to the information which covers legal ownership of the data [13]. Blockchain technology is multifaceted, differs multifariously from the existing information systems designs. It incorporates four

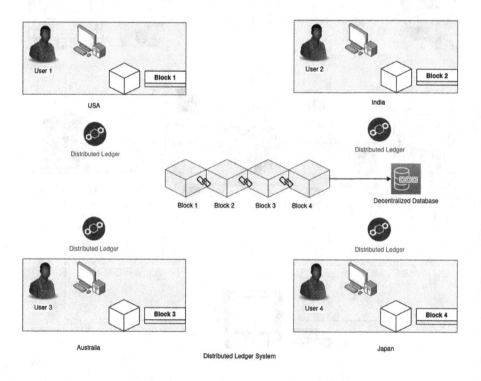

FIGURE 5.3 DLT system.

main features, i.e., non-localization (decentralization), safety, scrutinizing (tracking transaction route), and smart execution (via hash function) [14, 27]. Existing models of the network and blockchain mechanism differ from each other as the internet was designed to share any kind of information (not value) along with moving copies of original information so as to add to form a new knowledge base. In blockchains, information is encrypted and represented differently to distinguish transactional records in a secured manner. This feature gives it a new name as shared ledger and here data is safeguarded by providing a verifiable, time-stamped, and decoding key to access details of transactions, also auditing of information is done through encrypting-decrypting data.

In a p2p network, other blocks do perform verification/auditing on the joining of a new block and checks out everything, scanning whether changes are being done or not. In this technology, all the blocks make use of consensus to get a wise decision, assuring other blocks whether entry of the newly joined block is valid or not. On account of exorbitant measure of certainty present in the blockchain sequence, any new tempering occurred to the block will be depicted and will eventually lead to withdrawal of access by other blocks in this wide integrated network. To secure changes in the block, there is a need to get acknowledgment from the other blocks of the chain to enable alterations, needed to redo proof of the work, so it is very mind-boggling to temper the block-details [4].

The distributed network of computers provides versatile services which in turn help in rendering upper-hand to the blockchain automation which is significant in overcoming problems of load sharing and syncing data. For an efficient blockchain technology, distributed computing technology facilitates encrypting of information in graceful representation in a way to carry delicate information like intelligence data, medical information, business-managing decisions, transaction conducted, government documents, food-related details or electoral data, and many more.

5.7 ANALYSIS OF BLOCKCHAIN IMPLEMENTATION WITH BIG DATA

Blockchain as well as big data are contrasting approaches, harmonious to each other as these technologies can work simultaneously. Blockchain technology ensures safety of the data and enables encryption techniques to enhance the integrity among them. On the other hand, big data mechanisms take care of the alignment of data of all kinds and store statistics allied to data like type-of-data, data-quantity, size, velocity, and so on, necessary for data operations. While transacting, blockchain ledger generates huge data so their dealings are interconnected. The fusion of big data mechanisms and the blockchain sequence is the need of the hour and the coordination between the two is so secure that cannot be forged [9, 11]. There are three basic things that need to be comprehended where blockchain interrupts big data analytics.

- Decentralization: The major impediment is the cost of integrating big data analytics into an existing infrastructure [28]. With the advent of blockchain allied techniques, data analytics tools become more accessible that too for

a longer period of time because decentralization of the technology is done as per the need.
- Sharing and monetization: Data, gold of the new era is the most important set required to get information in today's world. The blend of concepts from big data and the blockchain has given ideas to share and monetize information for greater benefits [29]. By incorporating these phenomena business outcomes will face positive growth and there will be a surge in efficient customer dealings, negotiation, decision-making, etc.
- Interchange of data: There are multiple platforms currently available for exchanging data for instance Dock platform. It helps to manage the working environment of different job profiles and removes the idea of traversing multiple job sites hence providing relief to the working professionals.

Data exchanging through Dock is fully verified and is then consolidated to put it on the blockchain network. This certification will enable professionals to correlate things to form depth understanding about other steps.

The blockchain eradicates numerous risks like under-reporting, discrete data, tracing locations, and many more. Blockchain sequence is a new way to perform transactions and retains nascent form. Through innovation this technology can be revolutionized which will further supplement advancements [30]. According to the (Barrett) there is around 75% of the data which is unused and needs a suitable way to deal with it. To remove this conundrum, blockchain sequence is used in coordination with big data. This facilitates data security, hack-proofing, decentralization, and proper encryption while exchanging information within parties. The big data market will reach $230 billion by 2025 and collaboration with blockchain will improve smart spending. This advancement in services like transportation, healthcare, logistics media, and retail won't affect the pocket of the public and will remain affordable ("Big Data Market Worth $229.4 Billion by 2025 – Exclusive Report by MarketsandMarkets™"). Due to the eradication of disarranged ledger parts from the system more than 9/10 of total percentage of uncertainty is resolved and approximately half of the total percentage in effectiveness is sought through good execution speed. And more or less a quarter percentage of amelioration is achieved in public grievance redressal through swift processing of instructions by the system. Almost three-fourths of the total percentage is spent in a way to harmonize transactions, enhancing commerce to progress further (Aslam).

5.8 WAY FORWARD

A smart-city approach is an umbrella-term and multifaceted approach. Creating jobs and tax revenue in a smart city is conducive to a healthy economy. In order to attain sustainability, improving urban structures and town areas are inevitable. Future generations can benefit from the idea of smart cities not just in the short term, but also in the long term. Better use of space, cleaner environment, less traffic, satisfied citizenry, and efficient services all contribute to a better quality of life in smart cities. Furthermore, smart cities leverage greater economic opportunities and stronger ties to their communities. Also with the inclusion of emerging technologies like blockchain, AI, ICT, IoT, etc., smart-city concept can be further enhanced and it will supplement services in the field like drone utility, space tech, smart agriculture with profitable agribusiness, defense sector, governance security, check on cyber-attacks and so on. The alignment of the blockchain concept within society brings about smart services which in turn progress to achieve sustainable goals. In total, making premises smart enough reduces the chances for encountering uncertainty that will support the progress of a nation and aid sustainable resource utilization. And new applications have been made available to eliminate conundrums like traffic congestion, easy and quick delivery, global warming, and digitization of the economy and thus satisfy the citizenry.

5.9 CONCLUSION

The benefits of transforming traditional cities into smarter cities are huge and it has been further enhanced by incorporating blockchain-ledger systems. It somehow simplified the complexity of the IT sector and provided solutions to eliminate deformities from the system. The advances of information technology are the key drivers behind the emergence of Smart Services which transform ordinary cities into a city of automation. The more and more involvement of blockchain technology in sectors (like FinTech, insurance, secured IoT, data storage, online transactions, and other allied areas) will reduce the chances to encounter issues like data theft, vulnerable credentials, online attacks (e.g., phishing, DDoS, ransomware attack). This is enabled because of the security factor of the blockchain technology. Also collaboration of emerging technologies into smart communities has improved online delivery, reduces chances of warfare, and rejuvenates the view of surveillance and aiding in forming a robust and flexible economy. New technological development and smart approach to handle challenges with appropriate protocols can be chosen as major tools. The smart-city concept is a panacea that integrates ICT and its citizenry and both information and creates easy ways of data exchanges between different fields in minimal time. On an average everyday world encounters at least 2.5 quintillion bytes of data but still this data is not solely from smart services (Marr). So if the entire world needs to be reshaped into a smart world, solutions need to be found through advancing technologies that can handle such pressure. With the growing possibilities that modernization due to blockchain technology can further give out new scope for building techniques like quantum technology, AI automation, nanotechnology, and so on to uniquely structure each city for the betterment.

REFERENCES

1. Blockchain, Nofer, M., Gomber, P., Hinz, O., & Schiereck, D., 20 March 2017.
2. "Blockchains: The great chain of being sure about things". *The Economist*. 31 October 2015. Archived from the original on 3 July 2016. Retrieved 18 June 2016. The technology behind BTC lets people who do not know or trust each other build a dependable ledger. This has implications far beyond the crypto currency.
3. Su, K., Li, J., & Fu, H. (September 2011). Smart city and the applications. In *2011 international conference on electronics, communications and control (ICECC)* (pp. 1028–1031). IEEE.

4. Treiblmaier, H., Rejeb, A., & Strebinger, A. (2020). Blockchain as a driver for smart city development: Application fields and a comprehensive research agenda. *Smart Cities*, *3*(3), 853–872.
5. Eckhoff, D., & Wagner, I. (2017). Privacy in the smart city—Applications, technologies, challenges, and solutions. *IEEE Communications Surveys & Tutorials*, *20*(1), 489–516.
6. Ahmad, M. O., Ahad, M. A., Alam, M. A., Siddiqui, F., & Casalino, G. (2021). Cyber-physical systems and smart cities in India: Opportunities, issues, and challenges. *Sensors*, *21*(22), 7714.
7. Li, S. (2018, August). Application of blockchain technology in smart city infrastructure. In *2018 IEEE international conference on smart internet of things (SmartIoT)* (pp. 276–2766). IEEE.
8. Hassani, A. H., Huang, X., & Silva, E. 3, 19 October 2018, Big-Crypto: Big Data, Blockchain and Cryptocurrency.
9. Fusing Big Data, Blockchain and Cryptocurrency, Their Individual and Combined Importance in the Digital Economy, By Hossein Hassani, Xu Huang, Emmanuel Sirimal Silva, Figure: The fusion of Blockchain, Big Data and Cryptocurrency, 2019.
10. Nam, K., Dutt, C. S., Chathoth, P., & Khan, M. S. (2021). Blockchain technology for smart city and smart tourism: Latest trends and challenges. *Asia Pacific Journal of Tourism Research*, *26*(4), 454–468.
11. Yu, H., Yang, Z., & Sinnott, R. O. (2018). Decentralized big data auditing for smart city environments leveraging blockchain technology. *IEEE Access*, *7*, 6288–6296.
12. Saberi, S., Kouhizadeh, M., Sarkis, J., & Shen, L. (2019). Blockchain technology and its relationships to sustainable supply chain management. *International Journal of Production Research*, *57*(7), 2117–2135. http://doi.org/10.1080/00207543.2018.1533261
13. Abbaspour, A. (June 2018). Potential Benefits of Blockchain Technology for Mining Industry: With a Case Study of Truck Dispatching System in Open Pit Mines. Institute for Mining and Special Civil Engineering, TU Bergakademie Freiberg, Germany.
14. Kaya Soylu, P., Okur, M., Çatıkkaş, Ö, & Altintig, Z. A. (2020). Long memory in the volatility of selected cryptocurrencies: Bitcoin, Ethereum and Ripple. *Journal of Risk and Financial Management*, *13*(6), 107.
15. Haque, A. B., Bhushan, B., & Dhiman, G. (2022). Conceptualizing smart city applications: Requirements, architecture, security issues, and emerging trends. *Expert Systems*, *39*(5), e12753.
16. Alnahari, M. S., & Ariaratnam, S. T. (2022). The application of blockchain technology to smart City infrastructure. *Smart Cities*, *5*(3), 979–993.
17. Balcerzak, A. P., Nica, E., Rogalska, E., Poliak, M., Klieštik, T., & Sabie, O. M. (2022). Blockchain technology and smart contracts in decentralized governance systems. *Administrative Sciences*, *12*(3), 96.
18. Kumar, N. M., Goel, S., & Mallick, P. K. (2018). Smart cities in India: Features, policies, current status, and challenges. In *2018 technologies for smart-city energy security and power (ICSESP)* (pp. 1–4).
19. Dameri, R. P. (2017). "Smart city implementation," progress in IS. In Smart City Implementation. Springer: Genoa, Italy.
20. Tsaih, R. H., & Hsu, C. C. (2018). Artificial intelligence in smart tourism: A conceptual framework. ICEB 2018 Proceedings (Guilin, China). 89. https://aisel.aisnet.org/iceb2018/89.
21. Zheng, Z., Xie, S., Dai, H.N., Chen, X., & Wang, H. (2017). Blockchain challenge and opportunities: A survey. *International Journal of Web and Grid Services*, *14*(14), 352–375.
22. Paolini, P., Di Blas, N., Copelli, S., & Mercalli, F. (September 2016). City4Age: Smart cities for health prevention. In *2016 IEEE international smart cities conference (ISC2)* (pp. 1–4). IEEE.

23. Antonopoulos, A. (20 February 2014). "Bitcoin security model: Trust by computation". *Radar.* O'Reilly. Archived from the original on 31 October 2016. Retrieved 19 November 2016.
24. Polansek, T. (2 May 2016). "CME, ICE prepare pricing data that could boost bitcoin". Reuters. Retrieved 3 May 2016.
25. Decker, C., & Wattenhofer, R. (2014). Bitcoin transaction malleability and MtGox. In European Symposium on Research in Computer Security. Springer, 313–326.
26. Andoni, M., Robu, V., Flynn, D., Abram, S., Geach, D., Jenkins, D., McCallum, P., & Peacock, A. (2019). Blockchain technology in the energy sector: A systematic review of challenges and opportunities, Renewable and Sustainable Energy Reviews, 100, 143–174. http://www.sciencedirect.com/science/article/pii/S1364032118307184
27. Xu, J. J. (2016). Are blockchains immune to all malicious attacks? Bentley University, Waltham, MA 02452, USA.
28. Tasatanattakool, P., & Techapanupreeda, C. (2018). Blockchain: Challenges and applications. 473–475. 1https://doi.org/0.1109/ICOIN.2018.8343163.
29. Ammous, S. (August 2016). The Center on Capitalism and Society Columbia University Working Paper No. 91, Blockchain technology: What is it good for? August 8, 2016.
30. Back, A., Corallo, M., Dashjr, L., Friedenback, M., Maxwell, G., Miller, A., Poelstra, A., Timon, J., & Wuille, P. (2014). Enabling Blockchain Innovations with Pegged Sidechains.

6 Comprehensive Review Recent Advancements and Applications of Cyber-Physical Systems for IoT Devices

Umesh Kumar Lilhore, Sarita Simaiya,
Martin Margala, Prasun Chakrabarti, and Atul Garg

6.1 INTRODUCTION

A cyber-physical system (CPS) is a dynamic situation that integrates cyber and physical space. A CPS is a feedback-loop mechanism that contains a collection of physical devices (sensors and actuators) controlled by computer-based algorithms. Service providers use CPS to demonstrate their products to customers for better understanding. It models various real-time applications such as automobiles, factories, health, agriculture, and monitoring, among others. The main objective of CPS is to enhance the adaptability, flexibility, performance, functionality, reliability, protection, and accessibility of large-scale systems to optimize their implementation [1]. The CPS is a revolutionary technology that combines our digital and physical worlds. It is closely linked with multiple virtual and fundamental components to meet customer requirements and quality. The CPS defines the cyber ecosystem as a technological era processed, transmitted, and controlled by software applications. Over time, the external environment has been monitored using numerous sensors and the Internet of Things (IoT) [2].

Consequently, the CPS contains technology, equipment, sensors, controllers, and embedded devices that interact with components and services. A framework is a sophisticated technology that links a network of instruments, actuators, and controllers to acquire, filter, calculate, and evaluate physical surroundings knowledge and implement the outcomes to the external surroundings [3]. The CPS is a form of technology intimately linked to IoT devices. It is a network-based heterogeneous distributed solution that incorporates a mathematical formalism with the IoT and a computational component to manage it. As a result, it highlights the dynamic interaction between virtual and real realities. CPS technological advancements are crucial to enhancing the quality of life more effectively than ever. Nevertheless, network security dangers are getting more serious [4].

In addition, the CPS is experiencing difficulties evaluating security risks formed by the continuously involved process, and additional security concerns are now being discovered and developed. It is not easy to implement and secure the CPS's safety and privacy with its diversity and the variance of its elements. The many

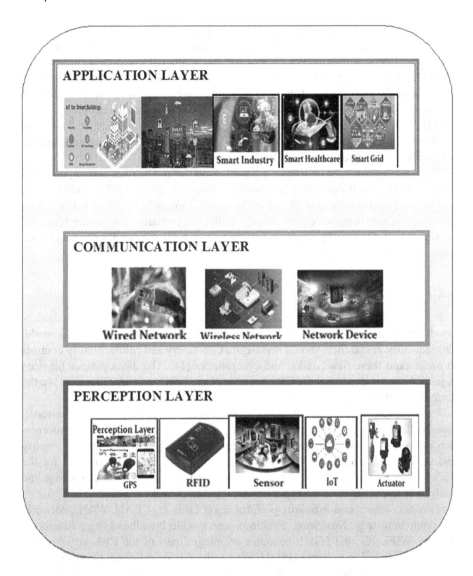

FIGURE 6.1 Overview of CPS.

CPS components and specific cyberattacks are also hard to ascertain, monitor, and investigate [5]. The first layer represents the perception layer, and the second layer represents the data transfer layer. The third layer represents the application layer [6], which constitutes the CPS, as displayed in Figure 6.1. The first layer is GPS, sensor, IoT, RFID, cameras, actuators, and identification [7].

CPS advancement is critical to enhancing living standards more effectively than ever, yet cybersecurity concerns are intense. In addition, the CPS has difficulty evaluating risks and vulnerabilities in complex connections, and additional security challenges have evolved [8]. Together with the CPS's elements, this diversity makes it challenging to verify the CPS's safety and confidentiality, discover, analyze, and investigate the CPA's various pieces, and concentrate on cyberattacks [9]. The interrelationship between CPS and IoT has remained a source of uncertainty inhibiting strong collaboration and information between the two areas. This chapter explains the beginnings of these aspects and analyzes various considerations. It presents a cohesive viewpoint that correlation improves security to encourage a common framework for sharing best practices, advancing common objectives, avoiding redundant work, lowering the propagation of conflicting specifications, and catalyzing exploration and advancement [10].

This chapter overviews current CPS for IoT platform advances and implementations. This chapter is organized into several categories. Section 6.1 introduces CPS and IoT and covers related work, while Section 6.3 explains CPS and IoT architecture. Section 6.4 discusses cybersecurity challenges and areas for further research, Section 6.5 discusses CPS applicability, Section 6.6 discusses CPS risks and remedies, and Section 6.7 concludes.

6.2 RELATED WORK

It is a significant issue that can potentially undermine the CPS's fundamentals by immediately compromising individual livelihoods in the physical world. Consequently, researchers should investigate CPS safety and confidentiality controls to understand these flaws, risks, and cyberattacks [11]. The discrepancies between organizational systems and the CPS in terms of the CPS's essential elements are the emphasis of this investigation.

It can capture audio, lighting, physical, pharmacological, temperature, electronic, biological, and geographical information. The sensors can produce factual information using node collaborations [12]. Consequently, the core network must identify and gather data, transfer it through the network layers, and function well for the site's IoT connections [13]. The information-sharing layer handles data exchange and analysis between sensors and applications [14].

This peer interaction is mainly performed via cable (e.g., LAN, WAN), networking equipment (e.g., Switching, Routing), and mobile broadband (e.g., Bluetooth, ZigBee, WiFi, 4G, and 5G). It becomes an integral part of the CPS, which might vary from internal to external [15]. Modern technologies are extraordinarily accessible and expensive since they can effectively handle and control large amounts of data across the Internet. A transceiver is also responsible for integrity and allows real-time streaming [16].

Based on the service, a distinct label may address the application layer. The "Supervisory Control and Data Acquisition" approach is the most common CPS, so it is employed in significant transportation systems projects like the Distributed Generation and the "industrial automation system" [17]. This layer receives an input transmission layer's source. It contains the instructions for the real sensing devices carried out and the instructions that could be employed in any section. In addition, data gathering from many sources, intelligence computation of enormous amounts of information, evaluation, and administration are planned [18]. Table 6.1 presents comparison of review on CPS risk, challenges, and security of IoT devices.

TABLE 6.1
Comparison of Review on CPS Risk, Challenges, and Security of IoT Devices

Citation	CPS Security Risk Factor	CPS Challenges
[19]	Computing capacity, storage capabilities, power capacity, scalable, and portability	Cost-benefit analysis, technical transfer, the lowest power protocols
[20]	Communication systems and data security	Flexibility, smart healthcare systems
[21]	Front-end monitors and instruments, communications, and the system's backend devices	Technological heterogeneity, energy-efficient approach, data sharing across boundaries
[22]	Energy-efficient approach, communication systems, and data security	Technology that works without wires, Self-organization skills, data and semantic interoperability, administration
[23]	The Internet of Things (IoT) links additional objects, security, and trustworthiness.	Specifications and accessibility are also concerns of complication, inconsistency, and integration.
[24]	Confidentiality has more decentralized network connections.	Internet service and energy consumption
[25]	Low protection on the application server for identification	Trustworthiness and various devices add to the system's functionality and communication issues.
[26]	Authenticate and confidentiality	Availability and expenses, system self-organization to ensure data security, and system failure.
[27]	The risk includes storing confidential material on a host machine while using a virtual private network when accessing a user's computer system.	Reliable connectivity and the availability of the required frequency band.
[28]	Whenever a protection system is built around a centralized controlled computer for processing, software, and storage systems, the risk of a system failure rises.	The expense of innovation. Insufficient organizational assistance. Leading to a shortage of well-trained personnel, IoT implementation is slow.
[29]	Security software, trust and safety, and physical security are all issues that need to be addressed. Hacking, denial of service (DoS), updates, viruses, login information assaults, and phishing are all examples of cybercrime.	IoT networks, solutions, and solutions are not well known.

6.3 CYBER-PHYSICAL SYSTEM AND IoT

Nowadays, computing devices are lightweight and portable, with physical components that allow tracking and control with proper action and real-time response. The terms "cyber-physical systems" and "Internet of Things" have different backgrounds but similar definitions relating to advances in incorporating technological transformation, including data connections and processing capacity, with external devices and objects [30]. The emergence of the particular phrase, "Internet of things," was imputed to "Kevin Ashton" in 1999. The IoT can be described as follows: "IoT aims to link the physical universe with the internet age" [31]. Similarly, CPS is defined as a "cognitive processing computer system that enhances learning in a product for specific tasks."

6.3.1 Physical Components of CPS

The critical components of CPS include:

- **One element of the cyber world:** The embedded devices share data via the Internet to handle the collected data automatically or through a human-machine interface (HMI). Computational units manage these embedded systems [32].
- **Physical constituents:** CPS systems have interacted with the physical environment through physical devices like sensors and actuators. Systems detect captured data in an electric form to evaluate or observe the physical world. Actuators in CPS are assumed to initiate behavior based on sensed data [33].
- **Physical components:** These are also known as plants, processes, and systems, and cyber components are known as computation, software, or code.

6.3.2 Architecture of Cyber-Physical System

CPAs' working principle consists of four stages: 1) monitoring, 2) networking, 3) computational processing, and 4) actuation. The cyber-world and physical space are CPS's two most prominent components [34]. Sensors in the physical environment continuously monitor the physical environment's objects, collect real-time data, and communicate this data to the cyber world. In cyberspace, decisions are made to perform specific actions depending on the raw data or information received. Moreover, the device's physical state changes from one state to another premised on the move.

As shown in Figure 6.2, the physical component consists of an actuator, computer system, and sensor. However, the cyber part consists of a communication network. CPSs are now based on an information processing system known as a "virtual system" embedded in a product. Sensors and devices in the computer system interact with the physical environment. An embedded system is a system that communicates with computer systems, actuation, and sensors to complete a given task. They do not live in isolation. They share their data with cloud technology via data transmission networks and the Internet, which processes the data collected from various embedded systems. As a result, a method of systems was formed [35, 36].

6.3.2.1 Configuration of CPS with IoT Devices

Despite numerous assumptions about the architectural layers of such a CPS, it operates on three levels. The accompanying devices classify the layers via the CPS architecture and even the underlying functional areas [37]. The CPS architectural style consists of three layers (see Figure 6.3):

a. Physical layer
b. Network layer, and
c. Application layer

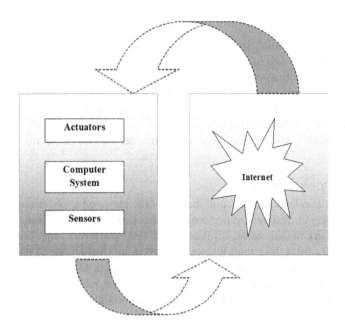

FIGURE 6.2 Physical components of CSP.

6.3.2.1.1 Physical Layer

The physical layer's commitment is to conduct more research using sensor devices. These sub-layers include the sensor, executor, and perception layers [38]. The sensor provides information from multiple physical devices, from analog to digital. The executor layer's purpose is to carry out command execution. The interpretation layer's functionality includes signal appreciation and data forwarding to the network layer. This layer comprises all the IoT devices required to observe the environment and deliver control behavior [39].

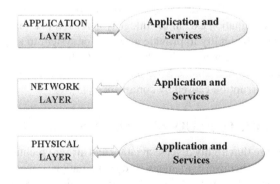

FIGURE 6.3 IoT architecture and CPS.

6.3.2.1.2 Network Layer

The network layer handles data transfer and communicates the application and physical layers. The connectivity infrastructure enables multiple components to connect, such as sensors, actuators, and other interconnected components. Heterogeneous communications systems may exist for wireless networks or TCP/IP networks. This layer manages CPAs' cost-effective, credible, and real-time communication. Bluetooth, ZigBee, WiFi, 4G, 5G, Local Area Networks (LAN), Wide Area Networks (WAN), switches, routers, and other network devices, are used in this layer [40].

6.3.2.1.3 Application Layer

In a CPS, the application layer combines all of the clustered applications. It refers to software packages designed to perform complex tasks, such as tracking and remote control, network management, databases, and HMI platforms. This layer works by collecting data from the network layer and executing instructions to physical units, known as actuators, to invoke the actions to be initiated. Many desktops and web-based interactions are available for administering and monitoring CPS functions. The user sends instructions to the control layer through this layer, which is executed [41, 42].

6.3.3 IMPACTS REGARDING CPS SECURITY IN IoT DEVICES

CPS ensures that almost all physical and cyberspace security goals are accomplished. The main objectives of CPS security are confidentiality, integrity, availability, and authenticity. These security targets are incredibly crucial for CPS, but they are insufficient. Physical processes will not be managed if a cyber-system is not usable. The results will be disastrous, especially for real-time operations [43].

- **Confidentiality:** It refers to the system's ability to prevent and avoid information leakage to intruders or the system itself from the outside [44].
- **Integrity:** The ability to protect information from being affected by an intruder. Data integrity is more valuable to the CPS than data confidentiality and privacy. The primary goal of a CPS domain is integrity, which would be achieved by avoiding, detecting, and preventing false attacks on innovative device information exchange. Integrity is related to a replay attack.
- **Availability:** The ability to provide a service or system to a specific group of participants. The availability objective of CPS seeks to guarantee that services are always available by avoiding computation, control, and communication corruption caused by connivance [45].

6.4 CYBERSECURITY ISSUES AND RESEARCH GAPS

It is critical to ensure that highly diverse aviation and aircraft technologies, including information security, firmware, memory, and computation, operate well. Research [46] underlined the significance of depending on following decades' CPS and highlighted how adaptable and introspective observable analytic approaches, critical to developing a realistic CPS, might be used as a foundation for organized adaptability. While using a communication methodology, this strategy radically shifts CPS's active surveillance, coordinating, and command domains. A high-level description of computer programming considerations was developed in the article [47].

The scheme is tied to the sophistication of the application development intention. In contrast, CPS technology is connected to the fundamental nature and kind of CPS to assure uniformity, assessment, architecture, implementation, and validation. The authors of the paper [48] discuss intelligent metering activities, network impact, and remote energy consoles, all of which directly affect the performance and maximum power supplied to the terminal. Moreover, researchers have discussed the matter and issues of CPS cybersecurity throughout this part because it is vital to enable power implementation and security architecture. Except for IT confidentiality, CPS confidentiality can have severe consequences for various companies and industries if it is breached. Consequently, previous articles' concerns and response plans [49, 50] should be considered in CPS cybersecurity.

6.5 CYBER-PHYSICAL SECURITY THREATS

Security issues to the CPS can result in severe significant violence. Then each component of the CPS can be physically or dynamically challenged. Enables devices and communications on the CPS are also exposed to other worldwide web assaults [51]. As a result, vulnerabilities to the area under consideration include challenges to different surroundings like sensing, actuation, and the ToT; concerns to the transport layer comprise communication vulnerabilities, computer viruses, and system errors.

6.5.1 CLASSIFICATION OF CPS SECURITY THREATS

The CPS security threats can be classified as given in Table 6.2.

TABLE 6.2
Classification of CPS Security Threats

CPS Security Threats		
Physical Layer Threats	Network Layer Threats	Application Layer Threats
Physical attack	Reply attack	SQL injection attack
Sensor hacking threat	Spoofing	Cross-site scripting
Cryptographic attack	Flooding attack	Buffer overflow
Fault injection threat	Routing attack	Malicious attack
Node reputation	Storage attack	SQL injection attack
Timing attack	Masquerade attack	Cross-site scripting
Node capture	MITM Attack	Buffer overflow
Eavesdropping	Selective forwarding	Malicious attack
Tag cloning	DoS attack	Buffer overflow
Jamming		Malicious attack

6.5.1.1 Threats to the Physical Layer

Devices in physical layers are typically located in an exterior and outdoor environment, which causes physical damage due to environmental change, mishandling, etc. This layer consists of sensors that identify objects and collect real-time data, along with actuators that measure changes in the air, temperature, motion, and so on [52]. Sensors are the primary target of attackers who want to design their sensors. More research is currently being conducted to find alternatives to these attacks [53]. The following are the major threats to the physical layer:

- **Physical attack:** A physical attack occurs when someone gains access to the devices. IoT devices are installed in environments where possible physical attacks may result in component damage/replacement. A biological attack disrupts, damages, or destroys IoT devices, equipment components, and data availability [54].
- **Sensor hacking risk:** The primary goal of sensor hacking in IoT devices is to generate incorrect data. Sensors are needed for safety-critical systems such as self-driving vehicles, drones, and medical equipment. Breaking these systems' safety can result in death or an accident. For these reasons, sensors are always designed as an alternative to failure or defects [55].
- **Cryptographic attacks:** CPS is entirely dependent on security data. This data is attacked, collected, and stored to be hacked into hardware devices. A cryptographic attack is an attempt to compromise the integrity of a cryptographic system by exploiting a fault in a cryptographic protocol, cipher, code, or essential management technique [56].
- **Fault injection threat:** Fault injection is a physical layer threat on the target device that involves injecting faults into the system to modify its intended functionality.
- **Node reputation:** Examples of node reputation attacks include node capture, false nodes, and node outages [57].
- **The complexity of test and analysis:** Privacy issues for the CPS can lead to numerous highly violent acts, so each element of the CPS is then addressed, whether physically or responsively. The equipment and interactions on the CPS are also vulnerable to various web-based attacks [51]. Consequently, threats to the region under examination involve difficulties in many environments, such as sensors, automation, and the IoT; risks to the network layer comprise networking weaknesses, malicious programs, and network faults.
- **The complexity of design and implementation:** Due to the abovementioned issues and constraints, the agile methodologies for the target CPS can be exceedingly complicated. Moreover, the CPS must meet many necessities imposed by different factors, such as the components, application logic, other development environments, programming languages and interface mechanisms, and external constraints [58, 59].

6.6 CYBER-PHYSICAL SYSTEM SECURITY SOLUTIONS

The cybersecurity architecture aims at protecting the complete atmosphere, comprising digital security, information systems, confidentiality, and the architectural and application surroundings. Security measures and blackhole attack classification algorithms leveraging software-defined networking (SDN) and industrial automation detection are examples of information system interconnect disobedience. The software-defined cyber-physical system (SD-CPS) was proposed by research [60] as a method and framework for addressing the CPS's implementation and design difficulties.

The SD-CPS leverages the SDN switching and microcontroller to incorporate the SDN when there is no SDN switching. As a result, the SD-CPS is compatible with current CPS implementations that do not incorporate SDN and can be used with them. It includes a wide range of research approaches for SDN deployment, design, enhancements, and alternative assessment methods [61] and established a novel detection technique in his investigation.

These information security strategies are depicted in Figure 6.4 and the CPS access control is divided into four categories: hardware security, connections identification, computer virus recognition, and infrastructure capacity [62].

Identification of computer viruses: Recently, several approaches for finding vulnerabilities have been developed [63]. These underlying mechanisms are utilized to identify hackers' software propagation in the CPS since they only depict identification on the Internet. The warning internet community networking worm identification method based on every Online social network's (OSN's) properties is suggested in research [64, 65]. The process utilizes the distance measurement system to determine user interactions by associating a "bait friend" with an actual OSN user population. Whenever there is proof, the identification system uses networks and local association algorithms to discriminate between harmful transmission and legitimate user information systems [66–70].

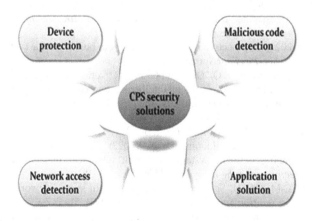

FIGURE 6.4 Cybersecurity schemes.

6.7 CONCLUSION

CPS is a revolutionary innovation that has laid the groundwork for many industrial innovations. It essentially involves interacting with various components and sensors, actuators, real-time generated data, process analysis, and a diverse range of interactive applications. This chapter intends to focus on the relevance of CPS security, laying the groundwork for future research. To provide insight into the new paradigm, we described the fundamental aspects of CPS and its challenges. Then, It articulated various threats to CPS and their solutions. The threats and CPS security solutions are inextricably linked. Most countries focus on this new technology to improve their quality of life.

Because CPS cybersecurity is a new sector that differentiates itself from the present online space, minimal research has been conducted. The CPS communication channel incorporates sensing, data kinds, actual statistical results, process analysis, and implementation interfaces. This chapter separates the dangers, remedies, and a CPS protection initiative highlights the CPS and offers an alternative to each problem. The CPS model and cybersecurity raised concerns and challenges and demonstrated the existing security market, including CPS-related research. The CPS private security protector appraised each level's dangers and remedies, suggesting future analysis prospects.

REFERENCES

1. Abbas, A. W., & Marwat, S. N. K. (2020). Scalable emulated framework for IoT devices in intelligent logistics-based cyber-physical systems: Bonded coverage and connectivity analysis. *IEEE Access: Practical Innovations, Open Solutions, 8,* 138350–138372.
2. Chaudhari, D. A., Chaudhari, D. A., Umamaheswari, E., & Umamaheswari, E. (2022). Adaptive deep rider LSTM-enabled objective functions for RPL routing in IoT applications. *International Journal of Information Security and Privacy, 16*(1), 172–189.
3. Adkane, D. R., Lilhore, U., & Taneja, A. (2016). Energy-efficient Reliable Route Selection (RRS) algorithm for improving MANET lifetime. *2016 International Conference on Communication and Electronics Systems (ICCES).*
4. Ammar, M., Crispo, B., & Tsudik, G. (2020). SIMPLE: A remote attestation approach for resource-constrained IoT devices. *2020 ACM/IEEE 11th International Conference on Cyber-Physical Systems (ICCPS).*
5. Ashraf, S. (2021). A proactive role of IoT devices in building smart cities. *Internet of Things and Cyber-Physical Systems, 1,* 8–13.
6. Azeez, N. A., Idiakose, S. O., Onyema, C. J., & Vyver, C. V. D. (2021). Cyberbullying detection in social networks: Artificial intelligence approach. *Journal of Cyber Security and Mobility.* https://doi.org/10.13052/jcsm2245-1439.1046
7. Battistoni, P., Sebillo, M., & Vitiello, G. (2021). An IoT-based mobile system for safety monitoring of Lone Workers. *IoT, 2*(3), 476–497.
8. Bouneb, Z. E. A., & Saidouni, D. E. (2022). Toward an IoT-based software-defined plumbing network system with fault tolerance. *International Journal of Hyperconnectivity and the Internet of Things, 6*(1), 1–18.
9. Burdescu, A.-M. (2021). Security of personal data in social networks. *International Journal of Information Security and Cybercrime, 10*(1), 51–58.
10. Celesti, A., Fazio, M., Longo, F., Merlino, G., & Puliafito, A. (2017). Secure registration and remote attestation of IoT devices joining the cloud: The Stack4Things case of study. In *Security and Privacy in Cyber-Physical Systems* (pp. 137–156). John Wiley & Sons, Ltd.

11. Chen, C.-M., & Liu, S. (2021). Improved secure and lightweight authentication scheme for next-generation IoT infrastructure. *Security and Communication Networks, 2021*, 1–13.
12. Chen, J., Tian, Y., Zhang, G., Cao, Z., Zhu, L., & Shi, D. (2021). IoT-enabled intelligent dynamic risk assessment of acute mountain sickness based on data from wearable devices. *2021 4th IEEE International Conference on Industrial Cyber-Physical Systems (ICPS)*.
13. Chowdhury, T. (2021). Towards reducing labeling efforts in IoT-based machine learning systems: PhD forum abstract. *Proceedings of the 20th International Conference on Information Processing in Sensor Networks (Co-Located with CPS-IoT Week 2021)*.
14. Dai, H., Shi, P., Huang, H., Chen, R., & Zhao, J. (2021). Towards trustworthy IoT: A blockchain-edge computing hybrid system with proof-of-contribution mechanism. *Security and Communication Networks, 2021*, 1–13.
15. Donati, L., Iotti, E., & Prati, A. (2021). A real-time approach for automatic food quality assessment based on shape analysis. *International Journal of Computational Intelligence and Applications, 20*(03), 2150019.
16. Dwivedi, S. K., Roy, P., Karda, C., Agrawal, S., & Amin, R. (2021). Blockchain-based Internet of things and industrial IoT: A comprehensive survey. *Security and Communication Networks, 2021*, 1–21.
17. El-Haouzi, H. B., Valette, E., Krings, B.-J., & Moniz, A. B. (2021a). Human and social dimensions in CPS & IoT-based automated production systems. In *Preprints*. https://doi.org/10.20944/preprints202107.0557.v1
18. El-Haouzi, H. B., Valette, E., Krings, B.-J., & Moniz, A. B. (2021b). Social dimensions in CPS & IoT-based automated production systems. *Societies (Basel, Switzerland), 11*(3), 98.
19. Fu, L., Ji, S., Lu, K., Liu, P., Zhang, X., Duan, Y., Zhang, Z., Chen, W., & Wu, Y. (2021). CPscan: Detecting bugs caused by code pruning in IoT kernels. *Proceedings of the 2021 ACM SIGSAC Conference on Computer and Communications Security*.
20. Garg, A., Lilhore, U. K., Ghosh, P., Prasad, D., & Simaiya, S. (2021). Machine learning-based model for prediction of student's performance in higher education. *2021 8th International Conference on Signal Processing and Integrated Networks (SPIN)*.
21. Goyal, N., Joshi, T., & Ram, M. (2021). Evaluating and improving a content delivery network (CDN) workflow using stochastic modeling. *Journal of Cyber Security and Mobility*. https://doi.org/10.13052/jcsm2245-1439.1043
22. Gressl, L., Steger, C., & Neffe, U. (2021). Design space exploration for secure IoT devices and cyber-physical systems. *ACM Transactions on Embedded Computing Systems, 20*(4), 1–24.
23. Guerar, M., Verderame, L., Merlo, A., Palmieri, F., Migliardi, M., & Vallerini, L. (2020). CirclePIN: A novel authentication mechanism for smartwatches to prevent unauthorized access to IoT devices. *ACM Transactions on Cyber-Physical Systems, 4*(3), 1–19.
24. Guleria, K., Prasad, D., Lilhore, U. K., & Simaiya, S. (2020). Asynchronous media access control protocols and cross-layer optimizations for wireless sensor networks: An energy-efficient perspective. *Journal of Computational and Theoretical Nanoscience, 17*(6), 2531–2538.
25. Guleria, K., Sharma, A., Lilhore, U. K., & Prasad, D. (2020). Breast Cancer prediction and classification using supervised learning techniques. *Journal of Computational and Theoretical Nanoscience, 17*(6), 2519–2522.
26. Haga, S., & Omote, K. (2021). IoT-based autonomous pay-as-you-go payment system with the contract wallet. *Security and Communication Networks, 2021*, 1–10.
27. Halba, K., Griffor, E., Lbath, A., & Dahbura, A. (2021). A framework for the composition of IoT and CPS capabilities. *2021 IEEE 45th Annual Computers, Software, and Applications Conference (COMPSAC)*.

28. Hassan, A., Prasad, D., Khurana, M., Lilhore, U. K., & Simaiya, S. (2021). Integration of Internet of things (IoT) in health care industry: An overview of benefits, challenges, and applications. In *Data Science and Innovations for Intelligent Systems* (pp. 165–180). CRC Press.
29. Jaiswal, R., Sahare, M., & Lilhore, U. (2018). Genetic approach-based bug triage for sequencing the instance and features. *2018 International Conference on Computer Communication and Informatics (ICCCI)*.
30. Jalbani, K. B., Jalbani, A. H., & Soomro, S. S. (2020). IoT security: To secure IoT devices with two-factor authentication by using a secure protocol. In *Industrial Internet of Things and Cyber-Physical Systems* (pp. 98–118). IGI Global.
31. Jan, Y., & Jozwiak, L. (2021). Quality-driven design of deep neural network accelerators for CPS and IoT applications. *2021 10th Mediterranean Conference on Embedded Computing (MECO)*.
32. Khalique, A., Alam, M. A., Khan, M. M., & Hussain, I. (2021). A security paradigm of WSN, IoT, and CPS: Challenges and solutions. In *Integration of WSNs into Internet of Things* (pp. 201–220). CRC Press.
33. Khapre, A., & Lilhore, U. (2016). MET OLSR – an energy effective OLSR based routing protocol. *International Journal of Computer Applications, 146*(13), 1–4.
34. Kolluru, K. K., Paniagua, C., van Deventer, J., Eliasson, J., Delsing, J., & DeLong, R. J. (2018). An AAA solution for securing industrial IoT devices using next-generation access control. *2018 IEEE Industrial Cyber-Physical Systems (ICPS)*.
35. Krishna, A., Le Pallec, M., Mateescu, R., & Salaün, G. (2022). Design and deployment of expressive and correct Web of Things applications. *ACM Transactions on Internet of Things, 3*(1), 1–30.
36. Kumari, A., Agrawal, N., & Lilhore, U. (2018). Clustering malicious spam in email systems using mass mailing. *2018 2nd International Conference on Inventive Systems and Control (ICISC)*.
37. Kumari, M., & Sharma, R. (2021). Comparative study of various forgery detection approach for image processing. *International Journal of Information Security and Cybercrime, 10*(1), 18–26.
38. Lemus-Prieto, F., Bermejo Martín, J. F., Gónzalez-Sánchez, J.-L., & Moreno Sánchez, E. (2021). CultivData: Application of IoT to the cultivation of agricultural data. *IoT, 2*(4), 564–589.
39. Li, S., Zhang, Q., Wu, X., Han, W., & Tian, Z. (2021). Attribution classification method of APT malware in IoT using machine learning techniques. *Security and Communication Networks, 2021*, 1–12.
40. Lilhore, U. K., Saurabh, P., & Verma, B. (2013). A new approach to overcome problem of congestion in wireless networks. In *Advances in Intelligent Systems and Computing* (pp. 499–506). Springer Berlin Heidelberg.
41. Lilhore, U. K., Simaiya, S., Guleria, K., & Prasad, D. (2020). An efficient load balancing method by using machine learning-based VM distribution and dynamic resource mapping. *Journal of Computational and Theoretical Nanoscience, 17*(6), 2545–2551.
42. Lilhore, U. K., Simaiya, S., Kaur, A., Prasad, D., Khurana, M., Verma, D. K., & Hassan, A. (2021). Impact of deep learning and machine learning in industry 4.0. In *Cyber-Physical, IoT, and Autonomous Systems in Industry 4.0* (pp. 179–197). CRC Press.
43. Lilhore, U. K., Simaiya, S., Prasad, D., & Guleria, K. (2020). A hybrid tumour detection and classification based on machine learning. *Journal of Computational and Theoretical Nanoscience, 17*(6), 2539–2544.
44. Liu, Z., Wu, L., Meng, W., Wang, H., & Wang, W. (2021). Accurate range query with privacy preservation for outsourced location-based service in IoT. *IEEE Internet of Things Journal, 8*(18), 14322–14337.

45. Lu, X., Fu, S., Jiang, C., & Lio, P. (2021). A fine-grained IoT data access control scheme combining attribute-based encryption and blockchain. *Security and Communication Networks*, *2021*, 1–13.
46. Malviya, D. K., & Lilhore, U. K. Department of IT, UIT BU, Bhopal, Madhya Pradesh, India (2018). Survey on security threats in cloud computing. *International Journal of Trend in Scientific Research and Development*, *3*(1), 1222–1226.
47. McCaig, M., Rezania, D., & Dara, R. (2022). Is the Internet of Things a helpful employee? An exploratory study of discourses of Canadian farmers. *Internet of Things*, *17*(100466), 100466.
48. Németh, L., Araya, G. Q., Regli, W., & Varró, A. (2021). A reference architecture for functional interoperability in robotics. *Proceedings of the Workshop on Design Automation for CPS and IoT*.
49. Nespoli, P., Díaz-López, D., & Gómez Mármol, F. (2021). Cyberprotection in IoT environments: A dynamic rule-based solution to defend smart devices. *Journal of Information Security and Applications*, *60*(102878), 102878.
50. Ngo, M. V., Luo, T., & Quek, T. Q. S. (2022). Adaptive anomaly detection for Internet of things in hierarchical edge computing: A contextual-bandit approach. *ACM Transactions on Internet of Things*, *3*(1), 1–23.
51. Pan, C., Xie, M., Han, S., Mao, Z.-H., & Hu, J. (2019). Modeling and optimization for self-powered non-volatile IoT edge devices with ultra-low harvesting power. *ACM Transactions on Cyber-Physical Systems*, *3*(3), 1–26.
52. Pandey, A., & Lilhore, U. K. (2017). An improved AES cryptosystem based genetic method on S-box, with 256 key sizes and 14-rounds. *International Journal of Advanced Engineering Research and Science*, *4*(3), 166–171.
53. Panoff, M., Dutta, R. G., Hu, Y., Yang, K., & Jin, Y. (2021). On sensor security in the era of IoT and CPS. *SN Computer Science*, *2*(1). https://doi.org/10.1007/s42979-020-00423-5
54. Panwar, N., Sharma, S., Wang, G., Mehrotra, S., Venkatasubramanian, N., Diallo, M. H., & Sani, A. A. (2022). IoT Notary : Attestable sensor data capture in IoT environments. *ACM Transactions on Internet of Things*, *3*(1), 1–30.
55. Verma, R. (2022). Smart city healthcare cyber physical system: characteristics, technologies and challenges. *Wireless personal communications*, *122*(2), 1413–1433.
56. Profentzas, C., Almgren, M., & Landsiedel, O. (2021). Performance of deep neural networks on low-power IoT devices. *Proceedings of the Workshop on Benchmarking Cyber-Physical Systems and Internet of Things*.
57. Pu, C., & Choo, K.-K. R. (2022). Lightweight Sybil attack detection in IoT based on bloom filter and physical unclonable function. *Computers & Security*, *113*(102541), 102541.
58. Raghuwanshi, V., & Lilhore, U. (2016). Neighbor trust algorithm (NTA) to protect VANET from denial of service attack (DoS). *International Journal of Computer Applications*, *140*(8), 8–12.
59. Rajan, C., Sharma, D., Samajdar, D. P., & Patel, J. (2020). Low power physical layer security solutions for IoT devices. In *Recent Advances in Security, Privacy, and Trust for Internet of Things (IoT) and Cyber-Physical Systems (CPS)* (pp. 229–248). Chapman and Hall/CRC.
60. Rúbio, E. M., Dionísio, R. P., & Torres, P. M. B. (2019). Industrial IoT devices and cyber-physical production systems: Review and use case. In *Innovation, Engineering, and Entrepreneurship* (pp. 292–298). Springer International Publishing.
61. Sandhu, M. M., Khalifa, S., Jurdak, R., & Portmann, M. (2021). Task scheduling for energy-harvesting-based IoT: A survey and critical analysis. *IEEE Internet of Things Journal*, *8*(18), 13825–13848.

62. Sandhya, C. P., & Manjith, B. C. (2021). Analysis of security issues, threats, and challenges in the cyber-physical system for IoT devices. *SSRN Electronic Journal.* https://doi.org/10.2139/ssrn.3882538
63. Schichl, H. (2021). Optimization problems in the power market - to verify or not to verify: Invited industrial talk at the 6th international workshop on symbolic-numeric methods for reasoning about CPS and IoT. *Electronic Proceedings in Theoretical Computer Science, 331.* https://doi.org/10.4204/eptcs.331.0.2
64. Sekaran, R., Kumar Munnangi, A., Rajeyyagari, S., Ramachandran, M., & Al-Turjman, F. (2021). Ant colony resource optimization for industrial IoT and CPS. *International Journal of Intelligent Systems, 22636.* https://doi.org/10.1002/int.22636
65. Shivaraman, N., Fittler, J., Ramanathan, S., Easwaran, A., & Steinhorst, S. (2020). WiP abstract: Mobility-based load balancing for IoT-enabled devices in smart grids. *2020 ACM/IEEE 11th International Conference on Cyber-Physical Systems (ICCPS).*
66. Shrivas, P., Lilhore, U., & Agrawal, N. (2017). Genetic approach-based image retrieval by using CCM and textual features. *2017 6th International Conference on Reliability, Infocom Technologies and Optimization (Trends and Future Directions) (ICRITO).*
67. Simaiya, S., Gautam, V., Lilhore, U. K., Garg, A., Ghosh, P., Trivedi, N. K., & Anand, A. (2021). EEPSA: Energy efficiency priority scheduling algorithm for cloud computing. *2021 2nd International Conference on Smart Electronics and Communication (ICOSEC).*
68. Simaiya, S., Lilhore, U. K., Sharma, S. K., Gupta, K., & Baggan, V. (2020). Blockchain: A new technology to enhance data security and privacy in Internet of things. *Journal of Computational and Theoretical Nanoscience, 17*(6), 2552–2556.
69. Singh, G., & Sharma, V. (2021). Cyber-security and its future challenges. *International Journal of Information Security and Cybercrime, 10*(1), 38–50.
70. Ali, R. A., Ali, E. S., Mokhtar, R. A., & Saeed, R. A. (2022). "Blockchain for IoT-based cyber-physical systems (CPS): Applications and challenges." In *Blockchain-Based Internet of Things* (pp. 81–111). Springer Singapore.

7 Deep Learning-Based Autonomous Driving and Cloud Traffic Management System for Smart City

Soujanya Syamal, Joyatee Datta, Srijita Basu, and Shuvendu Das

7.1 INTRODUCTION

The development of intelligent transport and traffic control system [1] is enhanced by rapid advancements in information technology and data communication. Intelligent transport control systems are a type of integrated systems that employ sophisticated technology. The traffic control system incorporates new parameters and proposals aimed at simplifying the system using an in-built cloud-based architecture [2] meant for autonomous driving. A self-driving car [3] is one that can move safely on its own while monitoring the objects in its environment. It collects data using a combination of sensors and human drivers, acquiring real-time results. Hundred percent self-driving with maximum efficiency is yet to be achieved but this model is the next step toward achieving autonomous driving. Several attempts have already been made to improve autonomous driving. This chapter presents a solution that aims to make autonomous driving more convenient and more competent in real-world scenarios. A Sense, Learn, and Act (SLA) model was proposed to improve autonomous driving in this chapter. The SLA model is inspired by modern-day, cutting-edge artificial intelligence (AI)-based systems that can continuously learn from their environment and react upon it. This study can be thought of as a quick introduction of a new smart-city concept [4] to achieve a safer and more comfortable transportation system.

The rest of the chapter has been organized as follows. Section 7.2 depicts the related work and highlights the novelty and importance of this work. Sections 7.3 and 7.4 describe the system architecture. Section 7.5 elaborates the workflow. Section 7.6 presents the result and finally, Section 7.7 concludes the chapter.

7.2 RELATED WORK

There are a good number of existing works on this topic. Most of the research is focusing on self-driving cars' algorithmic efficiency. Cloud-activated vehicles and autonomous vehicles are still two separated products in most of the surveyed works.

In this research an attempt has been made to combine these two aspects into one in order to make it a more efficient, dynamic, and responsive system.

In Ref. [5], a 3D-point cloud and SVM (Support Vector Machines) based system has been used to detect on-road signs and obstacles thus enabling autonomous driving. In Ref. [6], the different vital aspects of autonomous driving, viz., navigation system, path planning, environment perception, and car control, have been discussed. These aspects have been utilized in the proposed scheme. Moreover, real-time data has been fed into the deep learning model to evaluate the environment more precisely. The vehicle autonomy system will get more advanced and experienced by time and the amount of data it collects. Some other related work focuses on sensor to detect object or to plan the path [7], but in this study an attempt has been made to fuse the various sensor data to create a virtual world in vehicles' memory to understand the environment more efficiently. A driving simulator has been introduced in Ref. [8]. Here, the CARLA simulator [8] divides the driving task into different modules, viz., perception, planning, and continuous control, and uses reinforcement learning to train the system. It is to be noted that none of the existing works are implementing traffic management for autonomous vehicle. The traditional traffic management systems are quite old. So, this research was focused on cloud traffic management, in order to connect the autonomous vehicle with other vehicles and tried to create an ecosystem to make a well-managed traffic system.

Though some work has already been conducted on this topic, but an attempt has been made to handle some of the most overlooked problems in this area which reflects a scenario of huge connectivity of roads and number of vehicles at a time on the road, i.e., scalability. To address this issue, the system has been segregated into two parts, the local and global traffic management system, which addresses traffic management at granular levels of streets and utilizes that data to manage the traffic across the city [9].

7.3 SYSTEM ARCHITECTURE: AUTONOMOUS DRIVING

Autonomous driving in the cloud-connected traffic system has been proposed using a smart autonomous driving model, that is, the SLA model (see Figure 7.1). The car will sense the peripheral and learn through its system and will act as per its experience and learning model. Through deep learning algorithm, cars will learn from the sensor data and will act upon their experience.

7.3.1 COMPONENTS

7.3.1.1 Vehicle Navigation Framework

Two difficulties must be resolved during autonomous driving: the present position and the route from the source to the destination. The vehicle route framework is composed of various components such as global positioning system (GPS), route planning, and geographic data framework. These components are designed to provide the vehicle driver with unbiased route information. GPSs are used in vehicle navigation systems to get area data such as longitude and scope from satellites. The advanced guide data set and the street data provided by the area framework are

Deep Learning-Based Autonomous Driving

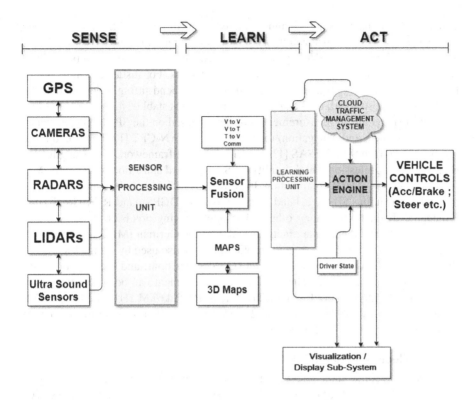

FIGURE 7.1 The SLA model.

among these data sets. They are generally utilized to organize the estimate of distinct model components, with the clever method of organizing calculations (like "Dijkstra calculation, Bellman-Ford calculation") being used to improve the way estimation is organized. A vehicle can find itself after estimating. With the data collected by oneself, the driving course can be modified to suit the model.

The navigation and positioning system's major function is to determine the vehicle's location, which is classified into three categories: relative area, outright area, and crossover area. A self-driving vehicle's current position is calculated by adding the moving distance and bearing to the prior location for the relative area. The vehicle's whirligig sensor and accelerometer acquire the vehicle's exact speed and sped-up speed in inertial navigation system (INS). The vehicle's total course point and speed can be established by coordinating this information (for example, accurate speed, sped-up speed). Incorporating the course point and speed can also be used to determine the vehicle's bearing and distance. The current vehicle area can be established by connecting it to the previous vehicle area. Nonetheless, due to vehicle vibration when traveling, a variation between the determined region and the true area is unavoidable. The outright area strategy is used to establish the vehicle's position, according to data obtained from the Location framework. A satellite-based positioning system, such as GPS, Global Navigation Satellite System (GLONASS) [10], Galileo [10], Bei-dou [10], and others, is a common example. Because of the climatic conditions, the satellite sign is slanted to the direction of the vehicle, which can produce a lot of confusion and inaccuracy in the area sign. The crossover area is a strategy utilized by self-driving car drivers to locate the ideal place to park. GPS/INS can also be used for routing and area applications. For instance, a vehicle can turn by following a map that was created by a line bend fitting. Even though the GPS signal is obstructed, the arrangement still remains stable. NovAtel [11] has two key GPS/IMU (Inertial Measurement Unit) items based on the SPAN (Synchronous Position, Attitude, and Navigation) innovation: the SPAN-CPT [12] integrated route framework and the SPAN-FSAS [13] fragmented route framework. The Length CPT uses a Novatel professional high-quality GPS board card and a fiber-optic gyro IMU from the German iMAR organization. Its response precision can be used in a variety of modes, including SBAs, L-band (Omnistar and CDGPS), and RTK (Real-Time Kinematic distinction), among others. The pitch-moving precision on this framework is $0.015°$, and the course exactness is $0.05°$. The German IMAR organization's [14] high-grade (HD) close-circle innovation IMU is also used by Range FSAS, and the gyro deviation acquired is less than 0.75 degrees/hour, and the accelerometer deviation is less than 1 mg. The blend route arrangement can be accomplished by combining it with NovAtel FlexPak6TM [11] or ProPak6TM [11]. The GNSS+INS framework has a yield speed of up to 200 Hz, while IMU-FSAS transmits inertial estimation data to the GNSS receiver.

7.3.1.2 Electronic Map (EM)

EM is used to guide data stockpiling, which for the most part incorporates topographical attributes, traffic data, building data, traffic signs, street offices, etc. Nowadays, the majority of the EMs used in autonomous vehicles are those designed for people. Currently, the high-definition (HD) map EM for self-driving vehicles has been released. In comparison to the conventional guide, the accuracy of outright arrangements of an HD map is superior from one point of view. For example, it is predicted that in the near future, drawing applications will be precise in millimeters, and that street traffic data components will be five key advancements toward the self-driving vehicle more lavish and precise. The dynamic layer, the unique layer, and the logical layer are the three layers that make up the HD map:

- In contrast to the traditional guide, the active layer includes HD street-level information (street shape, incline, ebb and flow, laying, bearing, and so on), path trait information (path type, path width, and so on), raised items, guardrail, trees, street edge types, side of the road milestones, and other massive objective data.
- On a regular basis, the dynamic layer will update traffic data from other cars and street sensors. The process of updating and supplementing is still continuing. This is the network integration community insight phase of the high-definition map's second phase.

The analytical layer aids in the training of self-driving vehicles by analyzing enormous amounts of data from human driving records. As a result, the HD map enters the third era of dynamic and control-enabled organization mix.

The action layer data is currently available on the advanced driver assistance systems (ADAS) [15] map, with a precision of 1–5 m. ASR ("Adaptive Speed Recommendation") [16] will analyze the width of the street, the number of routes, the general condition of the roadway, and other criteria while approaching a street turn to determine the vehicle's appropriate speed. Nowadays the high-definition map is basically ADAS level that may be used in self-driving cars with a level of autonomy of L2/L3 [17].

7.3.1.3 Map Coordinating

The vehicle's area is calculated using geological data from GPS or INS and guidance data from EM in map coordinating. During the estimation, the high-level melding procedure is utilized to intertwine the longitude and different directions data into the EM. From the pragmatic perspective, the yield of vehicle area ought to be exact and time-productive. In such a condition, finding a good approach to combine data from GPS and INS is a huge challenge. Because the satellite signal in GPS or INS can be lost at any time, a good information combining technique that incorporates data from the present region and course situation would vastly increase the precision, strength, and consistency of the system.

7.3.1.4 Universal Path Planning

The optimum path between the starting point and the destination is determined using Worldwide Path Planning. To combine the EM data and compute the optimum way, the most frequent methods of calculation are used, such as Dijkstra calculation [18], Bellman-Ford calculation [19], Floyd calculation [20], and heuristic calculation [21].

Even if the autonomous car is an important field of innovation that has been created and applied at the corporate level, there are still many issues to be solved in the future:

- The cost vs. precision: For the most part, the present area framework in an autonomous vehicle is based on the satellite area system; accordingly, it is important to decrease the expense later for huge scope business use, while simultaneously keeping up with the precision of the area.
- The trade-off between the area's exactness and speed: It is important to precisely find oneself driving a vehicle even in the high-velocity moving situation. Hence, acquiring a high-precision area under rapid conditions is a future exploration heading.

Creating an EM for an autonomous car that takes human individuality into account is the current need of the hour. Now coming to the proposed model, some perceptions are to be made in terms of the SLA model to input data from sensors and process it and act through learning the data from sensors.

7.3.2 Sense

Sense is the most important step for autonomous driving. When a human drives a vehicle, human has different organs to sense the environment to drive properly. But when it comes to autonomous driving, the vehicle should contain a similar medium to understand the environment and drive properly.

7.3.2.1 Environment Perception

The third module of a driving vehicle is climate discernment. To provide critical data to a vehicle's control decision, the vehicle must autonomously observe the general climate. Laser route, optical route, and radar route are three important climate discernment tools. Multi-sensors (for example, a laser sensor or a radar sensor) are sent to detect far-reaching data from the climate, which is then fused to perceive the climate during climate discernment. Radars are used for distance calculation.

7.3.2.2 Visual Perception and Object Recognition

For a self-driving vehicle, visual discernment is critical; for example, it is critical to detect traffic lights. The majority of traffic signals these days are built with human eyesight in mind; as a result, it is critical to recognize the traffic light. In addition, computer vision is used for determining location, route, motions, etc. The most confusing visual revelation, in particular, is how to ensure the calculation's unwavering quality and strength. The Camera is used to take image data to visualize the environment. In visual-discretion-based shrewd automobile routes, there are two basic improvement bearings. One can successfully visualize simultaneous localization and mapping (SLAM). Another example is a visual layout based on the knowledge of the collected image, which is processed using computer vision and AI. Following that, the person operating the automobile recreates the 3D environment in order to explore and perceive traffic lights, traffic signs, and stop lines. The SLAM is the focus of this chapter. The robot performs in an obscure region in an obscure climate, as depicted by the Hammer issue. During movement, the robot locates itself based on position assessment and sensor data and builds steady guidance at the same time. S_k is the information received from the sensors, M_{k1} is the closest guide at time $k1$, and R_k is the location of the self-driving car at time k. When the robot starts traveling from a concealed place in a secret climate, it uses position assessment data R_k and found information S_k to find itself and build the gradual guidance as it travels. The visual layout based on the knowledge of the collected image is processed using computer vision and AI. Following that, the person operating the automobile recreates the 3D environment in order to explore and perceive traffic lights, traffic signs, and stop lines. Also, the cloud will feed the connected cars with local and nearest traffic data to drive accordingly. Since, there is a cloud traffic system, and all the vehicles are separate nodes in the cloud, so one vehicle can easily know the relative position of the other vehicle established in the cloud as well as their relative speed. This can help to produce a more consistent flow of traffic. Two different boosting classifiers are used to complete the track arrangement: one on the states of the article at each casing in the track, and the other on the movement descriptors of the entire track. A discrete Bayes channel is used to connect these expectations. When the following divisions are correct, the classifier's precision is often about 98 percent. An aloof camera-based pipeline has been fostered for traffic signal state discovery, as well as in the cloud architecture there is a T to V communication, i.e., traffic to vehicle communication, in which the traffic sends its state whether it is red or not to the coming local vehicles toward the signal or crossing. When the vehicle receives the state, it can autonomously decide upon putting a brake or not. Thus, an extra secured layer is being added over the computer vision for detecting traffic signals. The vehicle already knows the traffic state and also knows when it will have to apply brake in signals. The traffic management follows this sequence: (1) First, the vehicle and the upcoming traffic post will connect and sync with each other. (2) Next, T to V communication will be established and the vehicle will receive the state of the signal. (3) Finally, if the state is RED, then the vehicle will decide to brake and will apply it accordingly.

7.3.2.3 Laser Perception and Unsupervised Calibration

In several aspects, the laser perception system is similar to a radar system. In laser discernment, a steady laser or laser beam is sent to the object, and the transmitter gets a reflected signal. The cloud information of the target point can be created by estimating the reflection time, reflection signal strength, shift of the activity recurrence, and then the testing object data, such as area (distance and point), shape (size), and state (speed and mentality), can be determined at that point. Though single-bar sensors can regularly be aligned without extraordinary trouble, inferring a precise adjustment for lasers with numerous concurrent bars has been a dreary and fundamentally harder test. A completely unaided way has been devised to deal with multi-bar laser adjustment, recuperating ideal boundaries for each shaft's direction and distance reaction work. The devised technique permits synchronous adjustment of many pillars, each with its own boundaries. Moreover, the sensor's extraneous posture has been recuperated.

An all-around model has been demonstrated using an iterative streamlining technique, the world steady alignment that is more precise than the industrial facility adjustment. Similar energy effort is used for the superfluous adjustment of the LIDAR's mounting [22] region on the vehicle, but in this case, an iterative usage of framework search is used to identify the best 6-DOF sensor present. The situation can be recovered to a resolution of 1 cm and the direction to a far finer granularity than is possible manually. Finally, an expectation-maximization algorithm has been employed to modify the forced settling re-turn esteems for each shaft, so accomplishing the requirement that pillars should typically agree on the (obscure) magnificence of surfaces in the climate. Following modification, the succeeding settlement maps are far more predictable.

In the case of conventional sign detection on road, the integrated cloud synchronization will serve the maximum purpose. And the camera system installed in the vehicle will do the rest of the job to detect real-time and relative objects, road signs, etc. The direction classifier is trained according to the map and navigation system. So according to the latest real-time location of the vehicle, it can easily find the signs.

7.3.3 LEARN

The *'learn'* parameters are as follows.

7.3.3.1 Sensor Fusion and Process

All the sensors' data will be processed to unique sets of data that can be combined and fused to make a revolutionary virtual environment in the vehicle's brain (CPU), and by the accumulated processed data the vehicle can learn in real time and can drive more efficiently, day by day.

7.3.3.2 Path and Direction Planning

The optimum path between the starting point and the destination is determined using Worldwide Path Planning. To combine the EM data and compute the optimum way, the most frequent methods of calculation are used, such as Dijkstra calculation, Bellman-Ford calculation, Floyd calculation, and heuristic calculation. The presented technique departs from this procedure and elevates speed and distance control to the level of arrangement. Furthermore, by combining the use of regulating and braking/speed increase, the calculation incorporates receptive obstacle aversion. In each arranging phase, the vehicle will follow the rest of the recently selected direction. While the basic method for selecting an expense utilitarian [23] is compatible with Bellman's rule of optimality, the directives restricting it should in any event be close to the independent vehicle's optimum traffic behavior. The Frenet-Serret [24] detailing and the moving edge approach is used to combine multiple sidelong and longitudinal expense functions for diverse errands, as well as to emulate human-like driving behavior.

7.3.3.3 Routing

The traditional method of routing is based on pre-prepared deep learning model based on satellite location services, which is very laggy and updated long time back. It is very difficult to navigate and drive through the existing map. So this research proposes an integrated cloud traffic management system which is combined with the autonomous algorithm of the vehicle to manage each vehicle efficiently. The traffic posts can be used as an IoT (Internet of Things) devices that can communicate with the local vehicles to manage their routing and to feed them with real-time data about the peripheral. Thus, autonomous car can navigate through the city and can route easily.

7.3.3.4 Lateral Motion

The beginning condition of the advancement is chosen as indicated by the recently decided direction in order to increase comfort and thus limit the squared jerk along the subsequent path. Quintic polynomials [25] can be used to meet this expense practically. Rather than identifying the optimal route and then modifying the coefficients to create a viable alternative, at first a direction set is established by mixing multiple end conditions. In the subsequent phase, the significant route with the lowest cost is selected. It's worth noting that, advancing along the ideal path, the left-over direction becomes the perfect arrangement in the next step. This approach overlooks the vehicle's non-holonomic feature at extreme low points, with the objective of dismissing the majority of the directions due to invalid ebbs and flows.

7.3.3.5 Longitudinal Movement

Rather than focusing on time or distance traveled in prior initiatives, the focus will be on comfort while still contributing to security at high speeds, because smooth improvements adapt much better to the traffic flow. As a result, the longitudinal jerk is likewise considered in the problem of improvement. A longitudinal direction set similar to parallel directions is created with the added expense useful because distance continuing, blending, and stopping at specific positions require direction, which depicts the transition from the current situation to a longitudinal, potentially moving, target position.

7.3.3.6 Consolidating Lateral and Longitudinal Curves
The combined expenditures of each direction are calculated as the weighted total in the final advance:

$$C_{tot} = K_{lat}\ C_{lat} + K_{lon}\ C_{lon} \qquad (7.1)$$

According to a basic understanding, it is sufficient for thruway direction age to identify all traffic situations as consolidating, following another vehicle, keeping a specific speed, pausing at one location, and all combinations thereof, which frequently collide.

7.3.4 ACT
Now after processing the real-time sensor data, the machine learns accordingly, and now it's time to drive the vehicle.

7.3.4.1 Vehicle Control
Heading directions of a car and speed are the most important aspects of vehicle control. Vehicle control functions primarily consist of gathering information about the vehicle's status and refining the vehicle's control method. The discernment module uses the EM data to estimate vehicle speed and bearing. The EM data comprises climate insight, vehicle state, driving objective, traffic guidelines, and driving information. The control goal is then computed by the vehicle control computation and supplied to the vehicle control framework. Finally, the vehicle control framework puts these instructions into action in order to manage the vehicle's bearing, speed, light, and horn, among other things. The control stage is the primary element of the driving vehicle that regulates the vehicle's numerous frameworks, including the vehicle's non-freezing stopping mechanism, vehicle drive against slip framework, vehicle electronic solidity program, car sensotronic brake control, electronic brake power conveyance, assister stopping mechanism, electronic control suspension, electric force directing framework, etc. The electronic control unit (ECU) and communication transport are the two main elements of the control stage. The control computation is mostly carried out by the ECU, while the correspondence transport comprehends the correspondence work between the ECU and mechanical elements.

7.3.4.2 Speed and Heading Direction
Vehicle self-status mostly comprises vehicle speed and direction information. For the speed, insight photoelectric code is used, whereas path discernment is done using

both photoelectric point code and potentiometer. As the system is synced with a real-time cloud, the vehicle gets other vehicles' statistics, such as speed, heading direction, etc., to control its speed and heading by calculating relative speed and direction with other vehicles locally.

7.3.4.3 Vehicle Control Technique

Proportional integral derivative (PID) calculation, or a more advanced version of PID computation, is widely used in vehicle control systems. In today's mechanical creation measure, the PID control calculation is the most well-known control calculation. The information terminated sign is $r(t)$, the criticism blunder signal is $e(t)$, the control signal decided by PID control [26] calculation is $u(t)$, and the present actual yield signal by the controlled article is $c(t)$. The PID calculation controls the aim by utilizing three limits and math chores, including extent, combination, and separation. The input circle's contribution to modifying the objective's condition is the difference between the control target and reality esteem. There are challenges in the exemplary PID calculation, such as complex boundary modifications, limited adaptability, and so on. Control accuracy is unusually low, particularly when the transmission framework is very nonlinear and the longitudinal impedance is excessively complex. The more complex PID calculations utilized by the most recent self-driving automobile can compensate for the shortcomings of good PID calculations.

7.4 SYSTEM ARCHITECTURE: CLOUD TRAFFIC MANAGEMENT

Cloud helps us maintain and synchronize a centralized database, along with providing the ability to set up real-time updates to the required systems, upholding the overall integrity of the product. In cloud architecture, several cloud components connect with one another using application programming interfaces (APIs), typically web services. These systems are easier to manage than typical monolithic systems because they keep complexity under control. The cloud architecture is made up of different layers, viz., Infrastructure as a Service (IaaS), Platform as a Service (PaaS), and Software as a Service (SaaS). The data is stored in the cloud database in multiple data centers across the globe. The cloud database structure differs from a relational database management system. It has several nodes for query services. This is required for the database to be accessed easily and completely using cloud services. Peer-to-peer communication is the best option for handling any query implemented by the user. Each node in the cloud database has a map to the data contained in each node. This map to the stored data makes it straightforward to access the information required to answer a specific query.

Regardless of whether nodes are added or withdrawn from the cluster, routing methods and algorithms that determine when to transport data chunks that are working together ensure that data is constantly available. When data requirements are copied to newly installed nodes, the data is served from the original location. The routing mechanisms start directing requests to the new node once it has an up-to-date version of the data.

Deep Learning-Based Autonomous Driving

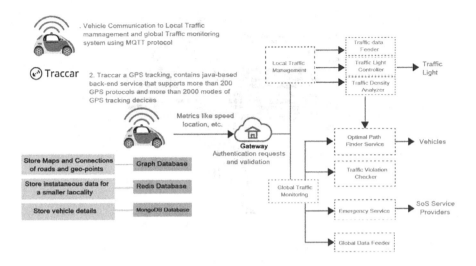

FIGURE 7.2 Cloud architecture.

7.4.1 Cloud-Based Vehicle and Traffic Control System

The proposed system focuses on the traffic light status, vehicle density, and relative speed on the road, for complete traffic management in a modern smart city. In order to ensure the smooth flow of the traffic, the traffic light management should be highly reliable and consistent and the traffic density and speed monitoring should be highly available, which can at times be inconsistent because of moving vehicles. As shown in Figure 7.2, the system is mainly divided into two parts, a local traffic management system, and a global traffic monitoring service. The key components of the service are small local Redis databases which are managed by a service for the locality-wise traffic control by the traffic signals to store the instantaneous data of a smaller number of cars. For traffic light control and other operations, in a small locality, a temporary storage in the form of a Redis database can be used. The global traffic monitoring service monitors the speed, location, engine status, and other records specific to each vehicle corresponding to the vehicle identity (ID) shared among a scalable MongoDB cluster and an SQL database.

It also stores the map and connection of roads and geo points using a graph database. In this system, each vehicle sends its data to a common service which in turn splits the data as required to the local and global traffic monitoring systems.

7.4.2 Local Traffic Management System

The local traffic management system consists of three services: the local data feeder, traffic density analyzer, and the traffic light controller.

The local data feeder service will accept each of the vehicle's speeds and locations on the corresponding road and store it in a Redis database for the region. The traffic density analyzer service will process the data of vehicles collected in a span

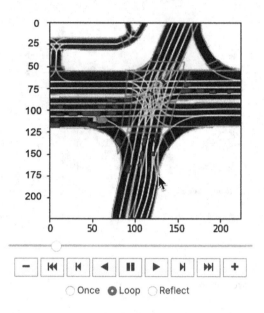

FIGURE 7.3 Cloud Traffic connection simulation.

of 2–3 minutes, and calculate the average traffic density and speed. According to the traffic density analyzer service, the current state of traffic signals of adjacent roads, and the presence of any emergency vehicle, the traffic light controller will decide the change in traffic signals. The car communicates to the local traffic management and global traffic monitoring using MQTT protocol.

7.4.3 Global Traffic Monitoring System

The global traffic monitoring system consists of a few services: the global data feeder, the traffic violation checker, the optimal pathfinder service, and the emergency service as shown in Figure 7.3. The global data feeder service will be responsible for collecting the data, like geo-location, average speed, vehicle ID, and type, and storing the required values in the database. The traffic violation checker service will monitor the conditions of the vehicle, like speed, traffic light violations, and make an entry of the corresponding fine in a database. The vehicles violating the rules can be identified and fined for the violations with a unique ID. The optimal pathfinder service will provide the optimal or least time-taking route for a vehicle using the local traffic density analyzer from the local traffic management system of each road on the way. In the case of any emergency, the emergency service will send alerts or messages by using a sensor, so the traffic system will change its state accordingly.

The important systems like traffic light management and optimal pathfinder algorithms cannot be hampered by data inconsistencies by any single vehicle as these services consider the data of all the vehicles in a particular area, before giving output.

7.5 WORKFLOW

The system architecture is designed in a way that it combined into a single workflow to work efficiently and in synchronized way to drive the car autonomously and manage the traffic simultaneously as depicted in Figure 7.4.

The vehicle will act as a node in the cloud system. When the vehicle starts, it gets connected to the cloud and the vehicle collects data from the cloud. After the navigation is being set, the car starts driving according to the SLA algorithm, by sensing the environment and obstacles. The cloud will feed the local traffic post device with the required information. Then the vehicle gets connected with the local traffic device and synchronization is done to navigate through the road. As every local vehicle is connected with the cloud traffic system, indirectly the traffic device is partially controlling the vehicle's speed and direction. The sensor is collecting the real-time data from the environment and the data is fed to the Sensor Processing Unit (SPU), where the data is processed and a unique virtual environment is generated and according to the data, the car understands the peripheral and drive through the road. Also, during the journey, the vehicle is continuously feeding the data to the cloud for a deep learning method to update the self-driving algorithm more efficiently and to synchronize other vehicles efficiently. The deep learning algorithm used to design the SLA model has been explained as follows.

The convolutional neural network's (CNN's) convolution layer performs feature extraction using several filters and is a key component of the design. The convolutional layer is made up of height and width filters [27, 28]. The convolution method transforms picture components of a specific size into a tiny array of numbers called kernels, from which the system extracts new representative information and processes it into a feature map. CNN is capable of recognizing things in images. Feature extraction determines CNN's capacity to recognize an item. Each thumbnail of the convolution results is used to create the feature map. The same filter is applied to all areas of each thumbnail. Each image's difference will be labelled as a unique

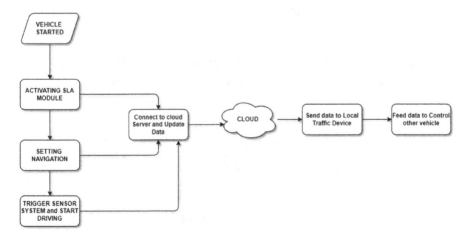

FIGURE 7.4 Workflow.

characteristic item. With a down-sampling operation, the pooling layer executes a sub-sampling technique to minimize input spatially (lower the amount of parameters). The convolution process is used to decrease the dimensionality of feature maps. Even if the number of parameters is reduced, the key information from the section is retained. A tiny array will be used to feed the feature map into deep neural networks. The flatten operation will turn the tiny array into a one-dimensional array. The probability of each image classification job done by the fully connected layers is the final neural network output. The CNN architecture is shown in Figure 7.5. Back propagation is used as an error correction to lessen the loss from the prior operation. It updates the weight values using an output error. The error is calculated by comparing the output of the forward propagation process (predicted value) to the actual value. The computation will adjust the weights based on the incorrect value. The method will be repeated until the error rate falls below a certain threshold.

The number of pooling layers can be changed to improve the model's performance. According to Lee et al., this is the most important aspect of CNN modelling [29].

The cameras' images, as well as the steering angle, are input into a CNN. After that, CNN creates a steering command prediction model. The prediction model is compared to the actual steering command model, and the CNN weights are changed to get the CNN output closer to the intended output. The weight modification is carried out through back propagation.

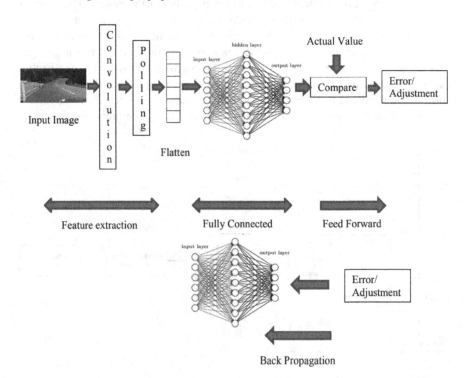

FIGURE 7.5 CNN deep learning layers.

Deep Learning-Based Autonomous Driving

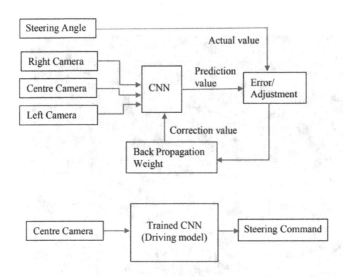

FIGURE 7.6 CNN model.

Following the completion of the training process, the system creates a steering model, which will be utilized for driving commands based on a single-center camera, as illustrated in Figure 7.6.

7.6 RESULT

A simulation according to the proposed idea was created and an autonomous algorithm was developed and connected to a cloud architecture, to regulate the vehicle movement. The network is built on the NVIDIA model [30], and the simulation environment is the VizViewer self-driving vehicle simulator [31]. Figure 7.6 depicts the learning environment produced in the simulation research, whereas Figure 7.7 depicts the graphs of orientation and motions, also the path planning can be seen. In Figure 7.2, the simulator of cloud traffic management which connects the autonomous vehicle and other connected vehicles can be seen. Also, Udacity [32] self-driving car simulator has been used, where about 16,000 photos were collected, each labelled with the kind of road and driver behavior, such as on lane or turning. The picture size is 320×160 pixels and the simulation sampling rate is 10 frames per second. For the learning process to be as simple as possible, the image captured will be normalized by deleting the sky and other superfluous components of the picture. To increase the quality of the classification findings, some pre-processing of acquired data pictures includes rotation, horizontal flip, vertical flip, and RGB to YUV conversion. The training method used a 5×5 kernel for the first three convolution layers and a 3×3 kernel for the last three, with a maximum of 20 epochs and 2,000 samples per epoch. After around two or three laps of training, the simulator switches to autonomous mode, and the car runs independently and predicts every road situation based on the CNN model. The simulation is run on a laptop with an M1 Max CPU, M1 Max GPU, and 16 GB of RAM. It takes around one hour to complete the learning process.

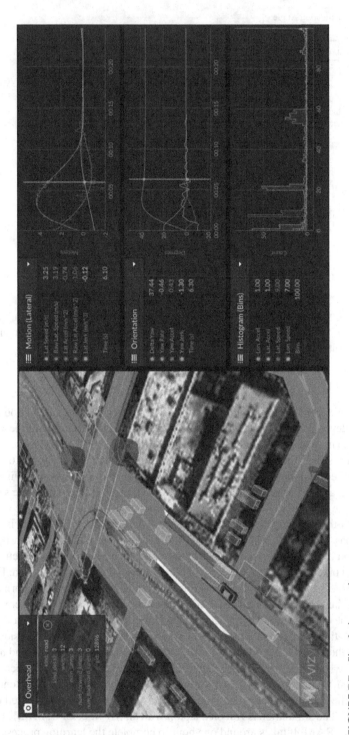

FIGURE 7.7 Simulation graph.

7.7 CONCLUSION

In a simulation environment utilizing the Udacity self-driving vehicle simulator the suggested approach of autonomous car employing CNN deep learning can operate smoothly without mistakes and is extremely stable without oscillation. CNN may use three cameras (left, center, and right) to learn about the road conditions and create a model for driving in autonomous mode using only the center camera.

An SLA model presents several solutions that aim to make driving safer and more competent in real-world scenarios. The car will sense the peripheral and learn through its system and will act as per its experience and learning model. Through deep learning, algorithm cars will learn from the sensor data and will act upon their experience. As the system is synced with a real-time cloud, the vehicle gets other vehicles' statistics, such as speed and heading direction. The important systems like traffic light management and optimal pathfinder algorithms cannot be hampered by data inconsistencies by any single vehicle as these services consider the data of all the vehicles in a particular area, before giving output. In the proposed system, the hardware-software interaction is designed in a way that will produce a better visual perception than the traditional autonomous car has. This Hardware system will be installed on the vehicle, so the processing and decision-making time will be very less. The traffic system is so systematic that it can be used as a solution of traffic jam in crowded cities. The next experiment will incorporate an obstruction in the simulation environment as well as a rain simulation with extra noise and distortion at the data picture set to improve performance.

REFERENCES

1. C. Liu and L. Ke, "Cloud assisted Internet of things intelligent transportation system and the traffic control system in the smart city," Journal of Control and Decision, 2022, 10, pp. 1–14.
2. Z. Khan and S.L. Kiani, "A cloud-based architecture for citizen services in smart cities," 2012 IEEE Fifth International Conference on Utility and Cloud Computing, November, 2012, pp. 315–320.
3. C. Badue, R. Guidolini, R.V. Carneiro, P. Azevedo, V.B. Cardoso, A. Forechi, L. Jesus, R. Berriel, T.M. Paixao, F. Mutz and L. de Paula Veronese, "Self-driving cars: A survey," Expert Systems With Applications, 2021, 165, p. 113816.
4. J. Winkowska, D. Szpilko and S. Pejić, "Smart city concept in the light of the literature review," Engineering Management in Production and Services, 2019, 11(2), pp. 70–86.
5. P.A. Raktrakulthum and C. Netramai, "Vehicle classification in congested traffic based on 3D point cloud using SVM and KNN," 2017 9th International Conference on Information Technology and Electrical Engineering (ICITEE), October, 2017, pp. 1–6
6. J. Levinson, J. Askeland, J. Becker, J. Dolson, D. Held, S. Kammel, J.Z. Kolter, D. Langer, O. Pink, V. Pratt and M. Sokolsky, "Towards fully autonomous driving: Systems and algorithms," 2011 IEEE Intelligent Vehicles Symposium (IV), June, 2011, pp. 163–168.
7. J. Zhao, B. Liang and Q. Chen, "The key technology toward the self-driving car," International Journal of Intelligent Unmanned Systems, 2018, 6(1), pp. 2–20.
8. A. Dosovitskiy, G. Ros, F. Codevilla, A. Lopez and V. Koltun, "CARLA: An open urban driving simulator," Conference on Robot Learning, October, 2017, pp. 1–16. PMLR.

9. R.P. Dameri, "Searching for smart city definition: A comprehensive proposal," International Journal of Computers & Technology, 2013, 11(5), pp. 2544–2551.
10. B. Hofmann-Wellenhof, H. Lichtenegger and E. Wasle, "GNSS–global navigation satellite systems: GPS, GLONASS, Galileo, and more," Springer Science & Business Media, Vienna, 2007.
11. P.C. Fenton and J. Jones, "The theory and performance of Novatel Inc.'s vision correlator," Proceedings of the 18th International Technical Meeting of the Satellite Division of The Institute of Navigation (ION GNSS 2005), September, 2005, pp. 2178–2186.
12. D. Li and H. Gao, "A hardware platform framework for an intelligent vehicle based on a driving brain," Engineering, 2018, 4(4), pp. 464–470.
13. H. Yu and J. Wang, "A new method to compute horizontal protection level based on vertical projection," 2016 IEEE Advanced Information Management, Communicates, Electronic and Automation Control Conference (IMCEC), October, 2016, pp. 901–905.
14. E.L.V. Hinüber, C. Reimer, T. Schneider and M. Stock, "INS/GNSS integration for aerobatic flight applications and aircraft motion surveying," Sensors, 2017, 17(5), p. 941.
15. R. Matthaei, G. Bagschik and M. Maurer, "Map-relative localization in lane-level maps for ADAS and autonomous driving," 2014 IEEE Intelligent Vehicles Symposium Proceedings, June, 2014, pp. 49–55.
16. B. Thomas, J. Lowenau, S. Durekovic and H.U. Otto, "The ActMAP-FeedMAP framework A basis for in-vehicle ADAS application improvement," 2008 IEEE Intelligent Vehicles Symposium, June, 2008, pp. 263–268.
17. Á. Takács, D.A. Drexler, P. Galambos, I.J. Rudas and T. Haidegger, "Assessment and standardization of autonomous vehicles," 2018 IEEE 22nd International Conference on Intelligent Engineering Systems (INES), June, 2018, pp. 000185–000192.
18. L. Parungao, F. Hein and W. Lim, "Dijkstra algorithm based intelligent path planning with topological map and wireless communication," ARPN Journal of Engineering and Applied Sciences, 2018, 13(8), pp. 2753–2763.
19. Y. Mo, S. Dasgupta and J. Beal, "Robustness of the adaptive Bellman–Ford algorithm: Global stability and ultimate bounds," IEEE Transactions on Automatic Control, 2019, 64(10), pp. 4121–4136.
20. J. Wang, Y. Sun, Z. Liu, P. Yang and T. Lin, "Route planning based on Floyd algorithm for intelligence transportation system," 2007 IEEE International Conference on Integration Technology, March, 2007, pp. 544–546.
21. S. Lin, and B.W. Kernighan, "An effective heuristic algorithm for the traveling-salesman problem," Operations Research, 1973, 21(2), pp. 498–516.
22. M. Jaboyedoff, T. Oppikofer, A. Abellán, M.H. Derron, A. Loye, R. Metzger and A. Pedrazzini, "Use of LIDAR in landslide investigations: A review," Natural Hazards, 2012, 61(1), pp. 5–28.
23. M.U. Shah, U. Rehman, F. Iqbal and H. Ilahi, "Exploring the human factors in moral dilemmas of autonomous vehicles," Personal and Ubiquitous Computing, 2022, 26(5), pp. 1321–1331.
24. M. Pilté, S. Bonnabel and F. Barbaresco, "Tracking the Frenet-Serret frame associated to a highly maneuvering target in 3D," 2017 IEEE 56th Annual Conference on Decision and Control (CDC), December, 2017, pp. 1969–1974.
25. A. Atabaigi, N. Nyamoradi and H.R. Zangeneh, "The number of limit cycles of a quintic polynomial system," Computers & Mathematics With Applications, 2009, 57(4), pp. 677–684.
26. M.A. Johnson and M.H. Moradi, "PID control," Springer-Verlag London Limited, London, UK, 2005.
27. R. Hecht-Nielsen, "Theory of the backpropagation neural network," Neural networks for perception, Academic Press, 1992, pp. 65–93.

28. C.Y. Lee, P.W. Gallagher and Z. Tu, "Generalizing pooling functions in convolutional neural networks: Mixed, gated, and tree," Artificial Intelligence and Statistics, PMLR, May, 2016, 51, pp. 464–472.
29. I. Sonata, Y. Heryadi, L. Lukas and A. Wibowo, "Autonomous car using CNN deep learning algorithm," Journal of Physics: Conference Series, IOP Publishing, April, 2021, Vol. 1869, No. 1, p. 012071.
30. G. Shainer, A. Ayoub, P. Lui, T. Liu, M. Kagan, C.R. Trott, G. Scantlen and P.S. Crozier, "The development of Mellanox/NVIDIA GPUDirect over InfiniBand—a new model for GPU to GPU communications," Computer Science-Research and Development, 2011, 26(3), pp. 267–273.
31. K.W. Brodlie, J. Brooke, M. Chen, D. Chisnall, C.J. Hughes, N.W. John, M.W. Jones, M. Riding, N. Roard, M. Turner and J. Wood, "Adaptive infrastructure for visual computing," Theory and Practice of Computer Graphics 2007, Eurographics UK Chapter Proceedings, Eurographics Association, 2007, pp. 147–156.
32. S. Du, H. Guo and A. Simpson, "Self-driving car steering angle prediction based on image recognition," arXiv preprint arXiv: 1912.05440, 2019.

8 Security and Privacy Challenges in IoT System Resolving Using Blockchain Technology

Gauri Shankar, Gaganpreet Kaur, and Sukhpreet Kaur Gill

8.1 INTRODUCTION

8.1.1 Internet of Things (IoT)

Internet-connected devices are replacing or integrating other technologies, making the system faster, more efficient, and smarter. These are only possible with an IoT network and devices. At the moment, IoT technology is being adopted in every industry sector as well as in everyday life. IoT is a combination of two technologies, the internet and smart devices, and its importance makes it a considerable future technology of the digital world [1]. IoT is being used as a key tool for the development of smart cities, in which all resources are connected to the internet through smart devices for better services. A continued evaluation of the IoT technology is required to understand the shared information of the user and their devices, as well as the architecture and advanced technology for that, so the information can be transmitted seamlessly [1]. These things are possible with the essential technologies used in the IoT, which are shown in Figure 8.1 and provide detailed discussion as follows.

8.1.1.1 Radio Frequency Identification (RFID)

RFID uses radio waves to capture data for automatic identification; for this process, two elements are used: the tag and the reader. The tag can store data more in the form of an Electronic Product Code (EPC) than a bar code. There are three types of tags: first, active tags, which have a self-contained battery to communicate; these can also contain different kinds of sensors to monitor different conditions. Applications of this type of tag are IT asset management, manufacturing, laboratories, hospitals, and remote sensing. The second type of tag is passive, which is powered by the reader and does not have its own internal battery [2]. Applications of this tag are in passports, item-level tracking, electronic tolls, and supply chains. And the third one is semi-passive; the tags use batteries to power the microchip at the time of communicating by drawing power from the reader [2].

Security and Privacy Challenges in IoT System

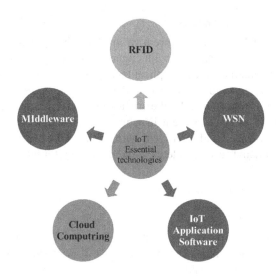

FIGURE 8.1 Essential technologies of IoT.

8.1.1.2 Wireless Sensor Networks (WSN)

WSN has devices connected to it that are used to monitor environmental or physical conditions and are integrated with the RFID to track the target objects or situations. In this case, a different type of network topology can be used to establish multi-hop communication. Recent research on low-powered small circuits for wireless communication is expanding the applications of WSN. Another application of WSN is to use it in tracking and maintenance systems that track the requirement of maintenance for a particular time period, which helps reduce the cost of irregular maintenance [3].

8.1.1.3 Middleware

The software application that helps to provide better communication and I/O between the connected devices is known as "middleware" in the IoT network. Any IoT system's developer uses feature abstraction to hide details of the technology used in the architecture [4]. This is involved in the facilitation of distributed computing architectures. The middleware is the best option to develop applications for complex IoT-based distributed computing architectures.

8.1.1.4 Cloud Computing

Cloud computing is an architecture of connected resources, such as storage servers, applications, and more, to share or access different kinds of systems, services, or any other connected infrastructures. In IoT infrastructure, cloud computing is most important as it generates huge amounts of data through its connected devices, and cloud computing is an ideal back-end solution for those data streams and processing them in unprecedented real time [5].

8.1.1.5 IoT Application Software

The development of IoT infrastructure also involved the application software that provides an interface for interaction with connected devices as well as its users in a reliable and robust manner. These applications are meant to ensure that the data is received and processed from the connected devices with the appropriate timing [6]. Data visualization is represented by the application software to the end user, which helps to understand the environment in which one is interacting. IoT devices are becoming more advanced and intelligent with application software, allowing them to monitor, identify, and solve problems without the need for human intervention.

8.1.2 FEATURES OF IoT

8.1.2.1 Connectivity

The most crucial part of IoT is connectivity. Without flawless communication among the interconnected components or devices, the IoT ecosystem (sensors, compute engines, data hubs, sensors and so on) cannot function correctly. Radio waves, Bluetooth, Li-Fi, and Wi-Fi are all options for connecting IoT devices [4, 5].

8.1.2.2 Analysing

The next stage is to analyse the data that is being collected and apply it to develop efficient business intelligence once all of the necessary objects have been connected. Extracting insight from the produced data is critical. A sensor, for example, creates data, but that data is useless unless it is correctly understood by humans [7].

8.1.2.3 Active Engagement

Passive engagement accounts, for a substantial portion of today's interactions with linked technologies. Multiple goods, cross-platform technology, and services collaborate on an active engagement basis through the IoT. In general, the usage of cloud computing in blockchain allows for active interactions among IoT components [5].

8.1.2.4 Scalability

More and more sectors are enabling the IoT zone every day. As a result, IoT systems should be able to manage significant expansion. The amount of data produced is a consequence is enormous, and it must be properly managed [8].

8.1.2.5 Artificial Intelligence

The IoT uses data collecting, artificial intelligence algorithms, and networked technology to make things like mobile phones, wearables, and automobiles smart and improve people's lives [7]. For example, apple watch has features that it detects if accident is happen to its owner and call an ambulance from nearby hospital and inform owner's relatives and friend from most near.

8.2 APPLICATIONS OF IoT

Some IoT applications are represented in Figure 8.2 and discussed as follows.

Security and Privacy Challenges in IoT System

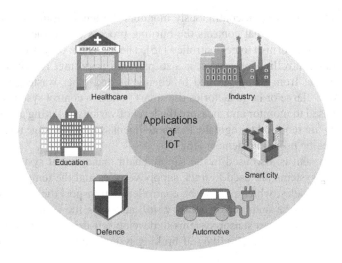

FIGURE 8.2 Applications of IoT.

8.2.1 IoT and Healthcare

The IoT system, also known as Internet of Medical Things (IoMT), is now playing an important role in the healthcare industry. There are many applications of IoT in the healthcare industry, such as tracking elderly patients' health at their homes, gathering health data from patients in hospitals, smart health monitoring, and alert devices for critical patients. Today, IoT is used in telemedicine systems in which surgeons can remotely perform surgery [9, 10]. Another critical application of the IoT system is mobile medical applications or wearable devices. These assist both users and physicians in tracking real-time healthcare data. At present, IoT is involved in different types of diagnosis processes with intelligent applications such as scanning and screening X-rays to easily identify the real issues. In Ref. [11], the author presents a system for monitoring the patient's or user's full body using a system of different kinds of sensors that are integrated with an IoT system and application software to collect real-time data, and the system is secured with an authentication process. The articles [12, 13] designed an IoT system to monitor and track a patient's health information that is integrated with blockchain to prevent unauthorized users.

8.2.2 IoT in Industry

Because IoT systems are constantly evolving the industry with new approaches, it is also known as the Industrial Internet of Things (IIoT). Currently, many different types of sensors are used in factories in the manufacturing process and quality testing. These sensors are connected through the IoT applications software, which collects data on production and quality testing to ensure quality and production speed [14]. Using an IoT system also benefits worker safety because there is no need for a worker to measure the temperature of the furnace; instead, a temperature sensor is attached to the furnish that continuously monitors the temperature and sends data to application software that controls the burning process of the fuel, allowing the temperature of the furnish to be controlled [15]. The IoT is not only useful for manufacturing and controlling processes; it also plays an important role in preventing the environment from being harmed by chemicals and other waste generated by factories. Many large or large factories use smart pollution control systems in which sensors are used to monitor and measure the level of various polluting elements, and because they are required to regulate, their intelligent systems manage to control the level of a specific polluting element [16].

The IoT system is also used in the supply chain management system, in which GPS tracking systems are used to track supply vehicles and the temperature of the inside of the vehicles, which need to be controlled to transport the product in fresh condition [17]. In the automotive industry, IoT advances the vehicular system by implementing specialized sensor systems to develop self-driving cars or improving security features of the vehicles like self-braking systems and driver monitoring systems. All of these are using advanced technology that integrates with IoT sensors or devices, and such computer vision systems track cars for self-driving mode as well as the driver's performance.

8.2.3 Education

Smart classes are one of the best examples of using IoT infrastructure in the education system. Using IoT devices, a simple campus turns into a smart campus, which means very little human effort will be required to operate the campus [18]. A smart campus can have the following facilities:

- Smart classrooms: in smart classrooms, smart boards are placed for teaching the learning process by collecting data from the class that teachers have written on the board and on the letters. They can be used inside the classroom as well as outside. There are many other sensors that will be used to control the noise and temperature of the classrooms, and the security camera is also connected to the IoT infrastructure and will be used to track the movement of students or faculty during the class. The smart ID card in classrooms is integrated with the smart attendance system that uses RFID to punch the attendance of students and faculty [19, 20].
- Smart laboratory: Equipping laboratories with sensors to prevent incidents is a smart decision, but using an IoT-based intelligence system in the laboratory to monitor the movement of students that can lead to an incident and alert the instructor is a smarter decision. At present, smart laboratories are equipped with surveillance devices that automatically ring the alarm when any incident happens and make a call to the related department of authority [19].
- Smart playground: Identifying the best in students involves monitoring and analysing their movements during sports activities, so the best use of an IoT in sports is the use of cameras and IoT application software to monitor and collect data from students and then analyse it for their improvement [19, 21].

In recent trends, the Raspberry Pi has played an important role in the development of low-cost IoT-based teaching and learning modules. Students are learning how to use the Raspberry PI and develop new tools for their classrooms [22, 23]. In addition, advancements in 5G technology will make the process of communication on devices connected to smart campuses faster than ever before [24].

8.2.4 E-Governance

E-governance is the management of all the things that can come under the authority of government, like resources, technologies, processes, and people. When the government is using digital technology to process the resources for their people, it is known as "e-governance," but when the IoT is involved in this process, it turns into "smart governance" [25]. Smart cities are the best example of smart governance, as in a smart city very few human efforts are required to handle the basic things needed to run the city, such as road traffic, government services, offices, and law and order. In smart cities, automatic traffic lights control the flow of traffic. A camera connected to the IoT system figures out how long it will take to clear the road based on how much traffic there is and lets the traffic that needs to go through [26].

At the moment, using government services is overly simple, as smart devices are used in government offices for both registration and service delivery. The best example of this is the Aadhar-integrated registration process, in which only an Aadhar number and a finger-print scanner are required to register for government services; there is no need to fill out any forms because the system retrieves data from the Aadhar storage server [27]. People are using government services like water connections, gas connections, and others, and they are able to pay for these services through their smartphones because of the Unified Payments Interface (UPI), which is a cloud-based service provided by the government that connects businesses and banks through a secure network [28, 29].

When it comes to security, law enforcement and the military are also employing smart devices for surveillance and apprehending criminals. In recent years, local police have used drones to monitor curfews in the city, as paths are small and not approachable in a short time. Security forces are doing the same thing, but in a more advanced way, as they use advanced surveillance systems linked to intelligent IoT-enabled infrastructure [30]. This type of defence technology is used to monitor the border for any kind of intrusion. In this, a drones or surveillance cameras are connected with the mainframe system through the internet or cloud computing systems that analyse the movement on the ground and in the air, and if any movement is unidentified or suspectable, immediately alert the defence personnel.

8.3 CYBER-ATTACKS ON IoT INFRASTRUCTURE

Cyber-attacks are categorized in:

- Software attacks
- Hardware attacks

8.3.1 ATTACKS ON IoT SOFTWARE

In this case, attackers are targeting the software applications of the IoT infrastructure in order to compromise the network. The attacker attempts to manipulate device firmware in several steps, the first of which is analysing the firmware update process and device components. Then attacker writes code during the firmware update process that can be executed to gain access to devices via debugging channels or the Joint Test Action Group (JTAG) (the JTAG provides standards and boundaries for scanning and debugging a printed circuit board).

Following successful access, the next step is to obtain responses from compromised devices via JTAG or serial console [31, 32]. Attackers carry out this action by employing software cracking tools, some of which are listed in Table 8.1.

8.3.2 ATTACKS ON IoT HARDWARE

The hardware attack is targeted at the hardware system of the IoT infrastructure, and the steps used to exploit vulnerabilities in the devices are the same as the software attacks. Some of the hardware attack tools are represented in Figure 8.3 and are used to exploit IoT infrastructure. In the term "hardware attacks" [31], they are furthermore divided into three types:

- Non-invasive attacks
- Semi-invasive attacks
- Fully invasive attacks

TABLE 8.1
Types of Software Attacks on IoT with Exploitation Tool

Types of Attack	Name of Tool
Web Testing	• ZAP
	• sqlmap
	• sslyze
	• Gobuster
Debugging	• GDB
	• OpenOCD
Bug Finder	• Flawfinder
	• Metasploit Framework
Binary Reversing	• IAD Pro
	• Radare2
	• Binaryaninja
Firmware Analysis	• Firmwalker: cpu_rec, binwalk
	• FAT (Firmware-Analysis-Toolkit)
	• FACT (Firmware Analysis and Comparison Tool)

Security and Privacy Challenges in IoT System

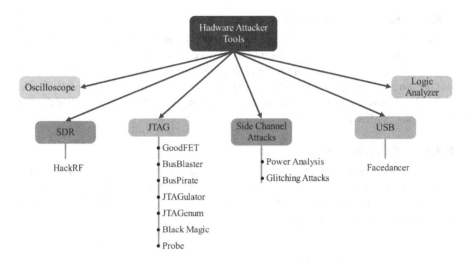

FIGURE 8.3 Types of hardware attacker tools.

8.3.2.1 Non-invasive Attacks

Attackers do not have direct access to device chips, but they can listen to IoT network traffic for a specific device and send forging data into JTAG or change the firmware update code to gain access, or send random or garbage data to devices to keep them busy working against their intended users [31, 33].

The non-intensive attacks are performed with a side channel attack, which is basically used to intercept network traffic, find vulnerabilities, and then execute exploitation. There are various types of attacks in this, such as:

- Timing attacks: These attacks target the device's computation timing, and if a vulnerability is discovered, it is used as a loophole in the device's security, specifically for authentication. Another thing that can be done by this attack is to monitor the missed cache and processed cache on the chip of the device and compare the difference; based on that, the attacker is able to find out the instructions written on the chip, access register values, or regenerate the memory content of the device [34].
- Hardware glitching: An attacker can manipulate the device's power supply and damage its chips by increasing or decreasing the voltage. In addition, attackers can control the clock speed of the device chip that controls the instructions, slowing down the chip's performance [35].
- Power analysis: The power analysis is done for the intercepted devices based on the processed data and the written instructions. As a result, the attackers can obtain information about the data being processed on the chip. These can be done with simple power analysis, differential power analysis, the EM radiation channel, or the acoustic channel [36].

8.3.2.2 Semi-invasive Attacks

This attack is performed on the chip physically. The attacker first analyses the target's location for the attack by disabling the device and emitting infrared light on the chip. In response to this action on the chip, photon emission is generated, and based on that, an attacker can analyse the chip and identify the target component to attack. Typically, this is used to break the device's encryption by changing the bit of the flip and employing leacher [31, 33].

8.3.2.3 Completely Invasive Attacks

This device attack requires expert knowledge and much more effort and cost than other attacks, but it is 100% successful. In this, the attacker modifies the chip by using an ion beam on it to generate a micro probing pattern of the chip, and the attacker can reconfigure it [31].

8.3.3 Types of Attacks on IoT Infrastructure

Based on the above categorization of attacks on IoT infrastructure, these attacks can be divided into different types for IoT devices as follows.

8.3.3.1 Privacy Attack

This attack is performed on the IoT network to gather information about an individual or group of users from the network. This attack can be used to steal an authenticated user's identity by regenerating any type of security access card or password. The attack can also be used to learn about the targeted user's financial situation and details by determining how many devices and expensive services he or she is using. As in IoT network infrastructure, a large number of users are connected, so if any other kind of attack is made, that may also risk the privacy of the users [37].

8.3.3.2 Device Attack

In the device attack, the main concept is to gain access to the IoT network by compromising any of the network devices. As a sensor, it will be compromised and used as a Trojan horse to send viruses on the IoT network that infect the whole network and provide access to attackers. This attack is very serious because if one device is infected, the whole network infrastructure will be affected. The process of detecting the infected device is time-consuming and costly, given the large number of connected devices. So, prevention is better than cure, and for that, device encryption and authentication may help [31, 37].

8.3.3.3 Network Attack

The attack is mainly targeting the communication performance of the IoT network through denial of service (DoS) attacks. It can cause communication delays or slow down network computation performance [32]. This attack is also known as "network availability." An example of this attack is to send false information into the processing unit of a banking IoT network, which would take a huge amount of time to verify the authenticity of the information. For the authenticated user, these types of attacks cause panic or slow communication in the IoT network. The communication process in the IoT network is critical, and even a minor incident can result in high-cost damage and security intrusion. Using encrypted data transmission and a decentralized data storage system, the attack can be avoided [32, 37].

8.3.3.4 Data Attack

The attack on data, transmitting between devices, and providing storage or utility in the IoT network to compromise the network weakens or exposes the infrastructure. It is more dangerous than a duplex data communication system. A simple example of this is a smart metre manipulation by a customer to save amounts on bills; the manipulation may harm the whole network. To prevent this, data integrity is a must inside the IoT network [32].

Based on these attacks on the IoT infrastructure and devices, it can identify the behaviour of different attacks and how much that are dangerous to the IoT infrastructure. Figure 8.4 shows the number of attacks reported by victims, and Figure 8.5 shows the losses reported from recent years. This report is generated based on some of the attacks listed in Table 8.2 and from the Internet Crime Complaint Center (IC3) only for the United States of America [42].

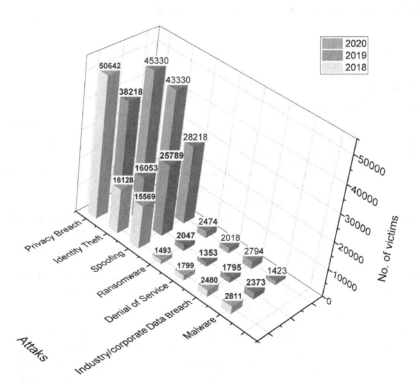

FIGURE 8.4 Types of attacks in recent years and number victims.

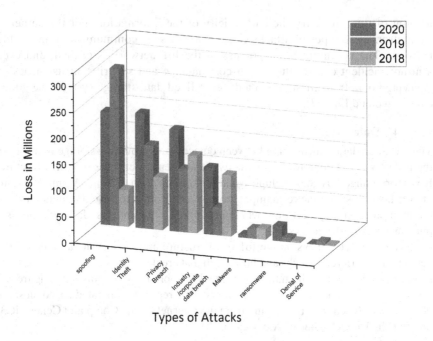

FIGURE 8.5 Types of attacks in recent years and losses in millions.

TABLE 8.2
Types of Attacks on IoT Architectures

Type of Attack	Description	Most Targeted Area of IoT Applications
Physical Attacks	Bypassing security and accessing the device physically, most of the time attack is performed through social engineering [32].	Defence, healthcare, industrial organizations.
Encryption Attacks	Attackers take advantage of weak encryption on device to compromise the network [32].	Defence, healthcare, industrial organizations, surveillance, supply chain, smart vehicle.
Distributed Denial-of-Service (DDoS)	Attacker creates a network of compromised systems and sent request to the server in intention to increase the traffic of network, results the legitimate user unable to use the services of sever which is under attack [38].	Service provider, supply chain, cloud storage provider, web services providers.
Firmware Hijacking	IoT devices which are not upgraded or updated can be targets of this attack. This attack is used as carrier to infect the whole IoT network [39].	Defence, healthcare, industrial organizations, surveillance, supply chain, service provider.

(Continued)

TABLE 8.2 *(Continued)*
Types of Attacks on IoT Architectures

Type of Attack	Description	Most Targeted Area of IoT Applications
Botnet	When attackers are compromised the IoT devices of the network and controlled them remotely from outside of the network. PBot malware and Mirai botnet are two types of attacks are mostly used [40].	Defence, healthcare, industrial organizations, surveillance, supply chain, service provider.
MIMT	This attack is used to monitor the network traffic between two system while data packet transmission. Attacker can also change the data and use malicious code to compromise whole network. This attack can be initiated by DNS spoofing, IP spoofing, SSL hijacking, HTTPS spoofing, Wi-Fi eavesdropping, Email hijacking, or stealing browser cookies [40].	Defence, healthcare, industrial organizations, surveillance.
Ransomware Attack	The IoT devices and the data on that will be encrypt by this attack by the attacker and damaged the access control system of authenticate user [41].	Defence, healthcare, industrial organizations, surveillance, supply chain, service provider.
Eavesdropping Attack	This is a process of interception of sensitive data in an unauthorized manner by the cybercriminal [40]. Attackers listing the data while IoT devices transmit data to server.	Defence, healthcare, industrial organizations, surveillance, service provider.
Privilege Escalation Attack	Attacker gain access through the system fault of the IoT devices and can modify the user access control also use compromised devices as a carrier for malicious code deployment on the IoT network [39].	Defence, healthcare, industrial organizations, surveillance, supply chain, service provider.
Brute-Force Password Attack	Attacker is trying to crack the encryption of the IoT network or any of the devices to gain access as authenticate user in the system [31, 40]. There are following types of brute-force attack: • Simple brute-force attacks • Dictionary attacks • Hybrid brute-force attacks • Reverse brute • force attacks • Credential stuffing	Defence, healthcare, industrial organizations, surveillance, education, automotive, supply chain, service provider.

8.4 SOLUTIONS WITH BLOCKCHAIN TECHNOLOGY

8.4.1 BLOCKCHAIN TECHNOLOGY

The blockchain is an immutable and distributed ledger technology in which participating nodes can anonymously write the transactions that are visible to all participating nodes of the ledger [43]. The basic elements needed to develop a

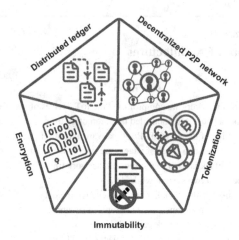

FIGURE 8.6 Key elements of blockchain.

blockchain are represented in Figure 8.6. Five basic elements of the blockchain are as follows:

1. Decentralized peer-to-peer network
2. Distributed ledger
3. Cryptography
    I. Cryptographic hashing
    II. Public key cryptography
    III. Digital signature
4. Immutability
5. Tokenization

8.4.1.1 Decentralized Peer-to-Peer Network

A network of connected nodes in a manner that each node can directly communicate with its neighbouring nodes without requirement of any central authority. In this system, a single node can be performed as server or client while responding or requesting over the network. A blockchain uses a decentralized network that provides decentralized management which ensures greater, fairer, and equal services to each node compared to a centralized network. Decentralization improves the communication as well as stability of the network. However, the rate of transaction is low [44]. Some advantages of decentralized network are as follows:

- It provides a thrustless environment where nodes do not require to be familiar with each other; every node maintains a copy of the information in their local system, and if they try to modify that copy, the modification will reflect on other connected nodes as well and be rejected based on the majority vote for verification of the information and validation of the modified information.
- Data is shared between nodes at runtime and are updated live, which can be viewed by shared nodes on the network, so if any changes are done by

Security and Privacy Challenges in IoT System

any node knowingly or unknowingly, other nodes can identify it and have option to correct it.
- Decentralization removes the single point of failure and corruption and provides fault tolerance distributed resources.

8.4.1.2 Distributed Ledger

A database that is synchronously shared with all the nodes connect in decentralized peer-to-peer network system. The ledger is publicly available to everyone on the network, which provides visibility over all the transactions processing on the network. Each node can have the same copy of the synchronized ledger, meaning every update is reflected to everyone. The distributed ledger solves the problem of the centralized authority to control ledger and prevent from corruption, cyber-attacks, fraud, as well as single point of failure [45]. Some advantages of the distributed ledger over the centralized one are as follows:

- This ensures that the ledger is prevented from alteration on any transaction by a single node; if it happens, other nodes also get acknowledged about it and majority of the nodes can remove the corrupted node without affecting the whole network.
- As the central authority is not required to process transaction in distributed ledger, cost is very less to process the transaction with high speed and is very efficient. Transaction processing is automated, making it 24/7 functional.
- In distributed ledger, auditing of the network is easy because of the flow of information; if the auditing report has any mistakes, then other nodes can oppose. This helps to prevent fraud or malicious act by the auditor.
- It provides fault tolerance against loss of the record as every node has the same copy.

In blockchain, a distributed ledger contains hash of IP addresses of all the nodes connected over the network that protects the privacy of the users. In place of storing the original data, the ledger stores a hash of the data that provides security and privacy.

8.4.1.3 Cryptography

In blockchain, the cryptographic protocols ensure the security and privacy of user and data. The main aim of these cryptographic protocol is based on CIA triad, which means confidentiality, integrity, and availability in the blockchain for users. In the process of cryptography, encryption and decryption function perform the key role. The process of achieving cipher text from plain text is known as encryption, and decryption means getting plain text from cipher text. The following cryptographic tools are used to achieve high level of security and privacy in the blockchain network [46].

8.4.1.3.1 Cryptographic Hashing

The cryptographic hash function can take any arbitrary size of data and produce a fixed length of output (259 bits in the case of blockchains). It is a one-way function, which means data is converted into hash and data cannot be generated from that hash. Cryptographic hashing function has the following properties that make it useful in the blockchain:

a. *Collision resistant:* this property states that two different values never have the same hash value such that X and Y if $X! = Y$ then $H(X)! = H(Y)$. But some old hashing algorithms like MD5 are found vulnerable to collision. In blockchain, Secure Hashing Algorithm SHA-265 is used.
b. *Hiding:* the hashing is a one-way function, which means that from the hash code, it is not possible to generate the original message, and this is achieved by using a secret value S selected from a probability distribution with a high mean entropy in such a way that $H(S||X)$ from this X will never be calculated.
c. *Puzzle friendly:* this states that for each n-bit output of a value Y, if k is selected from the high min-entropy, then it is impossible to find out the input value x such that $H(k || x) = Y$ in time significantly less than 2^n.

The use of hash functions in blockchain can be divided into six different categories, such as consensus (proof of work), in which a puzzle-friendly property is used. The second is address generation, and the third is to generate a hash of the previous block and add it to the Merkle tree. The fourth one is the message digest signature, the fifth is the pseudo-random number generator, and the last one is the wallet hash function.

8.4.1.3.2 Public Key Cryptography
There are two categories of cryptography.

1. *Symmetric:* Using the same key for encryption and decryption processes.
2. *Asymmetric:* Different keys are used for encryption and decryption process also known as public key cryptography.

Symmetric key cryptographic protocols are weaker these days so the blockchain is implemented in the public key cryptography [47]. A public key cryptography has the following properties:

- Keys are different and cannot be derived from each other, such that $k_1! = k_2$
- The sender and receiver can generate their key pair on their own system and share their public key only, but they must use the same public key cryptographic algorithm.
- Suppose the receiver's public key = $K_{R\ (pub)}$ and private key = $K_{R\ (pri)}$, then the process of public key cryptography is as shown in Figure 8.7.

Some of the public key cryptographies are Rivest–Shamir–Adleman (RSA), Elliptic Curve Cryptography (ECC), Diffie-Hellman, El-Gamal. In the first introduced blockchain, ECC was used to provide security and authenticity on the network.

8.4.1.3.3 Digital Signature
This is the tool that provides authenticity in the blockchain. Digital signatures sign a message that will be verified by the same signature algorithm to prove the authenticity of the message signature. It provides integrity, authentication, and non-repudiation

Security and Privacy Challenges in IoT System

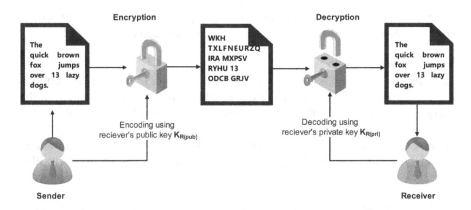

FIGURE 8.7 Public-key cryptography.

to the message. Through digital signature, signed personal identity cannot be identified whereas the signature identification is done by digital certificate of the signer [46]. This contains signer's info in hashed form. In digital signature, following three main processes are involved:

1. *(sk, pk): = generateKeys(keysize)* By this method, users generate a pair of keys in which *sk* is needed to be kept privately as it is the secret key and is used to sign messages. And *pk* is needed to be declared publicly for verification of the signature.
2. *sig: = sign(sk, message)* In this process, the secret key of signers is *sk* and a message is taken as input and a signed message is generated as output.
3. *isValid: = verify(pk, message, sig)* This is the process where one finds out whether the message is valid or not. Taking *pk, message* and *sig* as input, calculate *sk* of the signer. If the signature is matched, it is valid, otherwise it is invalid.

There are different types of signature algorithms such as RSA, ECDSA, EdDSA, ElGamal signature, and Schnorr signature. In blockchain, ECDSA is used a proof for authentication of messages communicated at the time of transaction.

ECDSA: The first blockchain bitcoin uses ECDSA's more standard version of ECC "secp256k1" which provides 128 bits of security to the message [48]. Some key elements of the ECDSA are as follows:

- Secret key: 256 bits
- Public key, compressed: 257 bits
- Public key, uncompressed: 512 bits
- Message to be signed: 256 bits
- Signature: 512 bits

It is very hard to breach the 128 bits of cryptographic security because it needs to perform large mathematical problems and requires high computation power. Most of the cryptographic algorithm takes years to be compromised. Also this uses trapdoor protocol which means that the private key is never recovered by the public key.

8.4.1.4 Immutability

The data of transaction is immutable when it is added to the block with performing consensus and done sufficient work to add it. This process is based on the energy consumption to compute the consensus for adding the block to the blockchain. After the addition of each new block, the complexity of the computation increases with each new block added to the chain, so the energy consumption also increases. As the complexity of the computation increases, immutability also increases, so the latest block has high immutability compared to the first block of the chain [49].

The immutability is based on the cryptographic algorithm used in blockchain and process of hashing, such that

$$Immutability = Blockchain\ hashing\ process + Cryptography.$$

Now how this immutability is archived in blockchain is as follows:

Every transaction on the chain is verified and timestamped at the time of adding on the chain and this is written on the associated block; all information of the block, including timestamp, is hashed by hashing function and added into the next block. This is performed like tree and the root of this tree is in latest block, and this tree known as the Merkle tree [50]. The process of adding hashes of previous blocks is consecutively applied to all blocks and this makes an immutable chain, because if any change happens in any block, all the hashes are changed and the blockchain will be invalid. Figure 8.8 shows an example of the Merkle tree.

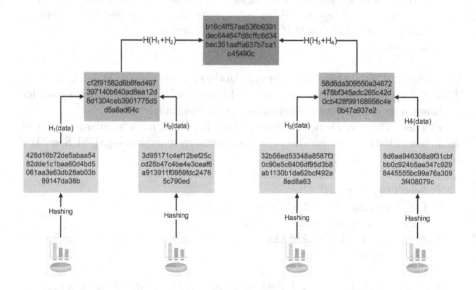

FIGURE 8.8 Merkle tree.

Security and Privacy Challenges in IoT System

8.4.1.5 Tokenization

The process of tokenization for any blockchain is to create a digital form of rights to transfer the ownership of the assets or some amount of assets. It can help to convert assets into token that can be further transferable. This token is familiarly known as cryptocurrency (bitcoin) in the case of bitcoin blockchain. There are two basic ways to create token, first is by using UTXO (spendable amount) for further transaction and other one is account based, meaning some amount of sender's wallet is transferred to receiver's wallet through smart contract (a self-executable program of agreement between parties in blockchain) [51].

There are following types of token present in blockchain:

- *Platform tokens*: this type of tokens is used in decentralizing app to govern their users
- *Utility tokens:* used for accessing services of the network such as pay transaction fees and consensus program
- *Fungible tokens:* replicable tokens, do not have uniqueness
- *Non-fungible tokens:* these are real-world things such as painting, tools, or any single piece of product that can be further divided, used as tokens

8.4.2 BASIC ARCHITECTURE OF BLOCKCHAIN

Definition: A decentralized peer-to-peer network of the ledger system, in which connected nodes can write the transactions to maintain their anonymity.

The structure of a simple blockchain (bitcoin) is presented as shown in Figure 8.9. Following are the elements of a single block:

- *Version:* it indicates which version of blockchain is present in the network. The size of this in the block is only 4 bytes. It also helps to detect downgraded hardware and software to prevent it from compromised in the network.

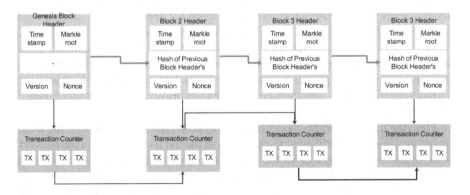

FIGURE 8.9 Blockchain architecture.

- *Timestamp*: it also takes 4 bytes; this is unique for every block which is used at the time of verification and validation of the block.
- *Hash of the previous block:* it takes 32 bytes inside the block. It is most crucial and the main element of the blockchain that stores information of the previous transaction in the form of hash.
- *Merkle root*: it is formed on the top of a tree that is generated by storing all the hash of previous transactions. It provides integrity to the transaction and make blockchain immutable.
- *Nonce*: it is a 4-byte filed that contains random numbers that are manipulated by miners to validate the block. Also used for the hashing the value of the block.

8.4.3 BLOCKCHAIN IN IOT INFRASTRUCTURE

IoT infrastructure consists of sensors and hardware devices and it has different layers discussed in Figure 8.9. From the layers of IoT infrastructure, it is clear that the lowest layer is sensing layer where sensing devices are planted to collect data and send to the upper layer in the network layers. The network layer defines how the devices are connected to each other and what type of protocol is used to establish communication between the nodes. This layer is the bridge between the data storage and other protocols governed in the middleware layer, where the Application Programming Interfaces (APIs) of the web tools or applications, database, and cloud storage are present. The security protocol is implemented in this layer to protect data loss. The data present in this layer is sent to the application layer for further processing. In application layer, there are different types of application that uses these data based on the requirements. These applications are run on advanced technologies such as machine learning or deep learning to analyse and give best results to application users.

Here is a case study of how blockchain technology and other tools were used to secure the IoT infrastructure in a smart city.

8.4.4 THE TRADITIONAL ARCHITECTURE OF SMART CITY

From Figures 8.10 and 8.11, it can be identified that the architecture of the IoT in smart cities is based on layers represented in Figure 8.9, and in Figure 8.12, entities of cloud storage are represented. The following attacks and vulnerabilities are possible on this architecture, which relies on cloud-centric storage and management. These attacks can be categorized against the features of blockchain, as listed in Table 8.3, that represent a brief explanation how blockchain is defensive against those attacks.

Based on the above countermeasures on the IoT infrastructure, the model depicted in Figure 8.13 is proposed with blockchain for smart city.

From the figure, it can be understood that the collected data is stored on the interplanetary file system (IPFS) and the hash of that the file is shared within the network with digital signature. IPs of all the nodes are hashed and the nodes are identified by their public key. Whenever any activity is performed by any node, the transaction details are written on the blockchain as a distributed ledger. The hash of data is shared with the prepossessing tool and that will fetch the data using hash, from

Security and Privacy Challenges in IoT System

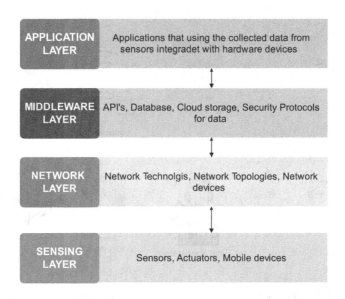

FIGURE 8.10 Layers in IoT infrastructure.

the IPFS. Then after pre-processed data is sent for analysis, the result and analysis data are sent to the IPFS. After getting a hash of that uploaded files, this hash is shared further to be used in IoT applications, which is fetched data from the IPFS using that hash. Each communication is digitally signed by the private key of sender and encrypted with the public key of the receiver. Each activity/communication is recorded on the blockchain with the addresses of senders and receivers, timestamp, and hash of the data. The process of communication between the nodes is shown in Figure 8.14.

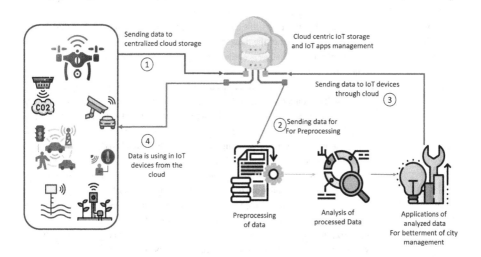

FIGURE 8.11 IoT infrastructure of smart city.

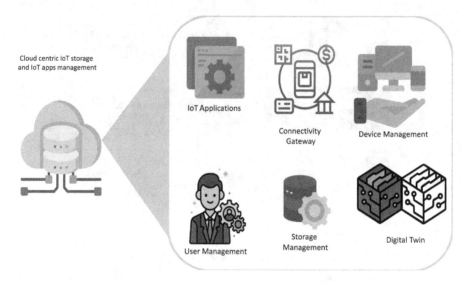

FIGURE 8.12 Entities of cloud storage in IoT infrastructure of smart city.

TABLE 8.3
Countermeasures against Attacks on the IoT Infrastructure, Using Blockchain Technology

Types of Attack	Prevention Using Blockchain
Privacy Attack	This attack targets the user's data on the network; however, blockchain uses SHA-256 to hash all the personal information of the users and protect their anonymity.
Timing Attacks	This attack targets the authentication protocol. As ECDSA is not easily deciphered, so this attack is not effective on blockchain network.
Hardware Glitching	If any device is attacked by this and the device behaves maliciously, then that device will no longer be part of the network.
Power Analysis	As the access to the devices is authenticated by the blockchain network, this attack is not effective.
Physical Attack	If any devices behave maliciously or do not follow the consensus of blockchain, then it will no longer be part of the network. So, even if anyone gets control on any device by physical attack, it cannot be active on the network after compromising.
Encryption Attacks	The ECDSA cryptographic protocol has 128-bit security key which is not easy to be broken at present by any computing tools through cryptanalysis.
DDoS	As the network is decentralized and distributed, the attack is not effective to the whole network and there is a large number of connected nodes present to provide data to client node apart from the nodes affected by this attack.
Firmware Hijacking	According to the version in the block, all the nodes are to be updated with the latest version of the software used by everyone. It prevents this attack on the blockchain network. Because if any node does not have the latest version of the network, it will not be treated as part of the network and will be abandoned.

(Continued)

TABLE 8.3 *(Continued)*
Countermeasures against Attacks on the IoT Infrastructure, Using Blockchain Technology

Types of Attack	Prevention Using Blockchain
Botnet	It is very hard to compromise any devices in blockchain because of the ECDSA cryptographic protocol that has 128-bit security key which is not easy to break at present by any computing tools.
MIMT	As the communication channels are encrypted with the ECDSA that provide security from network monitoring and having signature on data of sender provide integrity from being tampered while transmission.
Ransomware Attack	Because of distributed systems, data, or the ledger of hashed data is replicated to all the devices connected to the network. So, any system gets encrypted by ransomware attack. All the data can be recovered from any other device in the network very easily.
Eavesdropping Attack	This attack cannot affect as transmission channels are encrypted and transmitting data is authenticated with digital signature, so even if data can be listened and modified, it will be unauthenticated to the receiver.
Privilege Escalation Attack	This attack is prevented as the network is decentralized and every single communication between nodes/devices are encrypted and authenticated. So, even if the attacker gains access to a device it cannot change the access control. As each device performs the process in a distributed manner, the attacker cannot affect the whole network using a single compromised device.
Brute-Force Password Attack	In the network, ECDSA algorithm is used to encrypt communication and generate password, a set of key pair. This cryptographic algorithm is nearly impossible to be broken by key guessing attacks, because of its complexity in security bits and trapdoor property.

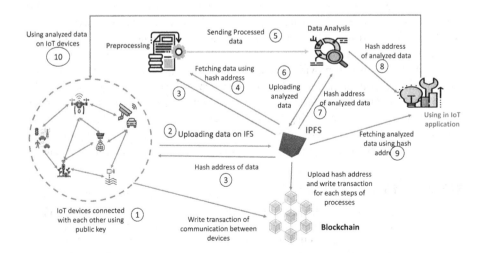

FIGURE 8.13 Blockchain-based IoT Infrastructure of smart city.

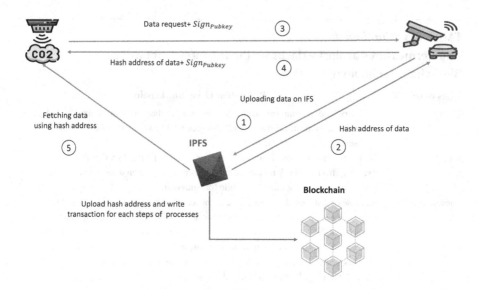

FIGURE 8.14 Communication between blockchain-based IoT infrastructures of smart city.

8.5 CONCLUSION

The IoT infrastructure is being implemented everywhere, paving the way for smart systems such as smart home appliances and smart cities. This type of adoption creates a smart environment in which a large amount of public and personal data is processed. It is mandatory to protect the privacy of the user's identity and secure the data from attackers. However, the traditional or basic IoT infrastructure is unable to fulfil this requirement. Integration of blockchain technology is the best option for accomplishing this. The properties of blockchain truly satisfy the IoT requirements to protect the privacy of the users and secure the data from the various kinds of attacks and vulnerabilities on the IoT infrastructure. The proposal of a smart city in the blockchain network demonstrates how blockchain works with IoT devices and provides security and privacy to users and their data. In that model, using IPFS, which is a decentralized and distributed file system, to store data, which provides hashes of data, as a result, it is possible to share a hash of data rather than the original data, which speeds up communication. The cryptographic algorithms that are being used in blockchain at present are secure, but not for long. As more powerful computing systems emerge, the strength of these cryptographic algorithms deteriorates. So, while developing an IoT infrastructure, especially for a long time, it is necessary to keep in mind that the design must be upgradable in the future.

REFERENCES

1. J. Gubbi, R. Buyya, S. Marusic, and M. Palaniswami, "Internet of things (IoT): A vision, architectural elements, and future directions," *Future Gener Comput Syst.*, vol. 29, no. 7, pp. 1645–1660, 2013. https://doi.org/10.1016/j.future.2013.01.010.
2. W. C. Tan, and M. S. Sidhu, "Review of RFID and IoT integration in supply chain management," *Oper. Res. Perspect.*, vol. 9, p. 100229, 2022. https://doi.org/10.1016/J.ORP.2022.100229.

3. A. Kumar, M. Zhao, K. J. Wong, Y. L. Guan, and P. H. J. Chong, "A comprehensive study of IoT and WSN MAC protocols: Research issues, challenges and opportunities," *IEEE Access*, vol. 6, pp. 76228–76262, 2018. https://doi.org/10.1109/ACCESS.2018.2883391.
4. M. A. Razzaque, M. Milojevic-Jevric, A. Palade, and S. Cla, "Middleware for internet of things: A survey," *IEEE Internet Things J.*, vol. 3, no. 1, pp. 70–95, 2016. https://doi.org/10.1109/JIOT.2015.2498900.
5. H. Tyagi, and R. Kumar, "Cloud computing for IoT," *Internet of Things (IoT): Concepts and Applications*, pp. 25–41, 2020. https://doi.org/10.1007/978-3-030-37468-6_2.
6. M. Pustišek, and A. Kos, "Approaches to front-end IoT application development for the Ethereum blockchain," *Procedia Comput. Sci.*, vol. 129, pp. 410–419, 2018. https://doi.org/10.1016/J.PROCS.2018.03.017.
7. M. Mohammadi, A. Al-Fuqaha, S. Sorour, and M. Guizani, "Deep learning for IoT big data and streaming analytics: A survey," *IEEE Commun. Surv. Tutor.*, vol. 20, no. 4, pp. 2923–2960, 2018. https://doi.org/10.1109/COMST.2018.2844341.
8. A. Gupta, R. Christie, and R. Manjula, "Scalability in Internet of Things: Features, techniques and research challenges," *Int. J. Comput. Intell. Res.*, vol. 13, no. 7, pp. 1617–1627, 2017. Accessed: Nov. 05, 2022. [Online]. Available: http://www.ripublication.com
9. W. Ying, P. Tran, and J. Wojtusiak. "From Wearable Device to OpenEMR: 5G Edge Centered Telemedicine and Decision Support System." HEALTHINF. 2022. https://doi.org/10.5220/0010837600003123.
10. S. Bhatla, G. Shankar, and M. Jalal, "A review of usefulness of virtual reality in treating autistic people," p. 050021, 2022. https://doi.org/10.1063/5.0109197.
11. K. H. Yeh, "A secure IoT-based healthcare system with body sensor networks," *IEEE Access*, vol. 4, pp. 10288–10299, 2016. https://doi.org/10.1109/ACCESS.2016.2638038.
12. S. Devi, Munisamy, et al. "Edge technology enabled IoT Blockchain-based health monitoring for chronically sick patients," *Next Generation of Internet of Things: Proceedings of ICNGIoT 2021.* Springer Singapore, 2021, https://doi.org/10.1007/978-981-16-0666-3_44.
13. N. V. Ravindhar, S. Sasikumar, N. Bharathiraja, and V. Kumar, "Secure integration of wireless sensor network with cloud using coded probable bluefish cryptosystem," *Jatit. org*, vol. 31, p. 24, 2022. Accessed: Jan. 13, 2023. [Online]. Available: http://www.jatit.org/volumes/Vol100No24/19Vol100No24.pdf
14. J. C. Cano, V. Berrios, B. Garcia, and C. K. Toh, "Evolution of IoT: An industry perspective," *IEEE Internet Things Mag.*, vol. 1, no. 2, pp. 12–17, 2019. https://doi.org/10.1109/iotm.2019.1900002.
15. F. Aslam, W. Aimin, M. Li, and K. U. Rehman, "Innovation in the era of IoT and industry 5.0: Absolute innovation management (AIM) framework," *Information (Switzerland)*, vol. 11, no. 2, Feb. 2020. https://doi.org/10.3390/info11020124.
16. M. A. Rahim, M. A. Rahman, M. M. Rahman, A. T. Asyhari, M. Z. A. Bhuiyan, and D. Ramasamy, "Evolution of IoT-enabled connectivity and applications in automotive industry: A review," *Veh. Commun.*, vol. 27. Elsevier Inc., Jan. 01, 2021. https://doi.org/10.1016/j.vehcom.2020.100285.
17. E. Manavalan, and K. Jayakrishna, "A review of Internet of Things (IoT) embedded sustainable supply chain for industry 4.0 requirements," *Comput. Ind. Eng.*, vol. 127, pp. 925–953, 2019. https://doi.org/10.1016/j.cie.2018.11.030.
18. H. K. Tripathy, S. Mishra, and K. Dash, "Significance of IoT in education domain," pp. 59–83, 2021. https://doi.org/10.1007/978-981-15-8621-7_6.
19. K. Zeeshan, and P. Neittaanmaki, "Internet of things enabling smart school: An overview," *2021 IEEE 18th International Conference on Smart Communities: Improving Quality of Life Using ICT, IoT and AI (HONET)*, Oct. 2021, pp. 152–156. https://doi.org/10.1109/HONET53078.2021.9615391.
20. N. K. Dewangan, P. Chandrakar, S. Kumari, and J. J. P. C. Rodrigues, "Enhanced privacy-preserving in student certificate management in blockchain and interplanetary

file system," *Multimed. Tools Appl.*, Sep. 2022. https://doi.org/10.1007/s11042-022-13915-8.
21. N. Bharathiraja, P. Padmaja, S. B. Rajeshwari, J. S. Kallimani, A. M. Buttar, and T. B. Lingaiah, "Elite oppositional farmland fertility optimization based node localization technique for wireless networks," *Wirel. Commun. Mob. Comput.*, vol. 2022, p. 5290028, 2022. https://doi.org/10.1155/2022/5290028.
22. S. Mahmood, S. Palaniappan, R. Hasan, K. U. Sarker, A. Abass, and P. M. Rajegowda, "Raspberry PI and role of IoT in education," *2019 4th MEC International Conference on Big Data and Smart City, ICBDSC 2019*, pp. 1–6, 2019. https://doi.org/10.1109/ICBDSC.2019.8645598.
23. T. Lei, Z. Cai, and L. Hua, "5G-oriented IoT coverage enhancement and physical education resource management," *Microprocess. Microsyst.*, vol. 80, Feb. 2021. https://doi.org/10.1016/j.micpro.2020.103346.
24. T. Lei, Z. Cai, and L. Hua, "5G-oriented IoT coverage enhancement and physical education resource management," *Microprocess. Microsyst.*, vol. 80, no. October, p. 103346, 2021. https://doi.org/10.1016/j.micpro.2020.103346.
25. V. Rohokale, R. Prasad, N. Prasad, and R. Prasad, "Interoperability, standardisation and governance in the era of internet of things (IoT)," *Internet of Things - Global Technological and Societal Trends from Smart Environments and Spaces to Green Ict*, pp. 257–285, 2022. https://doi.org/10.1201/9781003338604-11.
26. F. Al-Turjman, and J. P. Lemayian, "Intelligence, security, and vehicular sensor networks in internet of things (IoT)-enabled smart-cities: An overview," *Comput. Electr. Eng.*, vol. 87, Oct. 2020. https://doi.org/10.1016/j.compeleceng.2020.106776.
27. J. Shanmugapriyan, R. Parthasarathy, S. Sathish, and S. Prasanth, "Secure electronic transaction using AADHAAR based QR code and biometric authentication," *2022 International Conference on Communication, Computing and Internet of Things, IC3IoT 2022 - Proceedings*, 2022. https://doi.org/10.1109/IC3IOT53935.2022.9767978.
28. B. Ainapure, S. Khamparia, A. Varshney, and N. Baheti, "IoT enabled smart water ATM with digital payment," *Proceedings of 3rd International Conference on Intelligent Engineering and Management, ICIEM 2022*, pp. 189–194, 2022. https://doi.org/10.1109/ICIEM54221.2022.9853074.
29. A. Thiruneelakandan, G. Kaur, G. Vadnala, N. Bharathiraja, K. Pradeepa, and M. Retnadhas, "Measurement of oxygen content in water with purity through soft sensor model," *Meas.: Sens.*, vol. 24, p. 100589, 2022. https://doi.org/10.1016/j.measen.2022.100589.
30. M. Diwakar, K. Sharma, R. Dhaundiyal, S. Bawane, K. Joshi, and P. Singh, "A review on autonomous remote security and Mobile surveillance using internet of Things," *J. Phys. Conf. Ser.*, vol. 1854, no. 1, p. 012034, 2021. https://doi.org/10.1088/1742-6596/1854/1/012034.
31. C. Quast, "Common attacks on IoT devices: Why you can not win," in *OpenIoTSubmit Europe*, 2018.
32. J. Deogirikar, and A. Vidhate, "Security attacks in IoT: A survey," *Proceedings of the International Conference on IoT in Social, Mobile, Analytics and Cloud, I-SMAC 2017*, pp. 32–37, Oct. 2017. https://doi.org/10.1109/I-SMAC.2017.8058363.
33. M. S. Mispan, B. Halak, and M. Zwolinski, "A survey on the susceptibility of PUFs to invasive, semi-invasive and noninvasive attacks: Challenges and opportunities for future directions," vol. 30, no. 11, Mar. 2021. https://doi.org/10.1142/S0218126621300099.
34. S. Takarabt et al., "Cache-timing attacks still threaten IoT devices," *Lecture Notes in Computer Science (Including Subseries Lecture Notes in Artificial Intelligence and Lecture Notes in Bioinformatics)*, vol. 11445 LNCS, pp. 13–30, 2019. https://doi.org/10.1007/978-3-030-16458-4_2.
35. A. Gangolli, Q. H. Mahmoud, and A. Azim, "A systematic review of fault injection attacks on IoT systems," *Electronics*, vol. 11, no. 13, p. 2023, 2022. https://doi.org/10.3390/ELECTRONICS11132023.

36. J. Moon, I. Y. Jung, and J. H. Park, "IoT application protection against power analysis attack," *Comput. Electr. Eng.*, vol. 67, pp. 566–578, 2018. https://doi.org/10.1016/J.COMPELECENG.2018.02.030.
37. H. HaddadPajouh, A. Dehghantanha, R. M. Parizi, M. Aledhari, and H. Karimipour, "A survey on internet of things security: Requirements, challenges, and solutions," *Internet of Things (Netherlands)*, vol. 14, Jun. 2021. https://doi.org/10.1016/j.iot.2019.100129.
38. R. Khader, and D. Eleyan, "Survey of DoS/DDoS attacks in IoT," *Sustain. Eng. Innov.*, vol. 3, no. 1, pp. 23–28, 2021. https://doi.org/.37868/sei.v3i1.124.
39. M. Bettayeb, Q. Nasir, and M. A. Talib, "Firmware update attacks and security for IoT devices," *Proceedings of the ArabWIC 6th Annual International Conference Research Track on - ArabWIC 2019*, pp. 1–6, 2019. https://doi.org/10.1145/3333165.3333169.
40. G. Rajendran, R. S. Ragul Nivash, P. P. Parthy, and S. Balamurugan, "Modern security threats in the internet of things (IoT): Attacks and countermeasures," *Proceedings - International Carnahan Conference on Security Technology*, vol. 2019-October, Oct. 2019. https://doi.org/10.1109/CCST.2019.8888399.
41. M. Humayun, N. Z. Jhanjhi, A. Alsayat, and V. Ponnusamy, "Internet of things and ransomware: Evolution, mitigation and prevention," *Egypt. Inform. J.*, vol. 22, no. 1, pp. 105–117, 2021. https://doi.org/10.1016/J.EIJ.2020.05.003.
42. Federal Bureau of Investigation. "2020 Internet crime report." (2021), https://www.ic3.gov/Media/PDF/AnnualReport/2020_IC3Report.pdf.
43. N. K. Dewangan, and P. Chandrakar, "Patient-centric token-based healthcare blockchain implementation using secure internet of medical things," *IEEE Trans. Comput. Soc. Syst.*, 2022. https://doi.org/10.1109/TCSS.2022.3194872.
44. R. Schollmeier, "A definition of peer-to-peer networking for the classification of peer-to-peer architectures and applications," *Proceedings - 1st International Conference on Peer-to-Peer Computing, P2P 2001*, pp. 101–102, 2001. https://doi.org/10.1109/P2P.2001.990434.
45. D. Burkhardt, M. Werling, and H. Lasi, "Distributed ledger," *2018 IEEE International Conference on Engineering, Technology and Innovation, ICE/ITMC 2018 - Proceedings*, Aug. 2018. https://doi.org/10.1109/ICE.2018.8436299.
46. L. Wang, X. Shen, J. Li, J. Shao, and Y. Yang, "Cryptographic primitives in blockchains," *J. Netw. Comput. Appl.*, vol. 127, pp. 43–58, 2019. https://doi.org/10.1016/J.JNCA.2018.11.003.
47. R. Kumar, and R. Tripathi, "Secure healthcare framework using blockchain and public key cryptography," *Adv. Inf. Secur.*, vol. 79, pp. 185–202, 2020. https://doi.org/10.1007/978-3-030-38181-3_10.
48. S. J. Basha, V. S. Veesam, T. Ammannamma, S. Navudu, and M. V. V. Subrahmanyam, "Security enhancement of digital signatures for blockchain using EdDSA algorithm," *Proceedings of the 3rd International Conference on Intelligent Communication Technologies and Virtual Mobile Networks, ICICV 2021*, pp. 274–278, Feb. 2021. https://doi.org/10.1109/ICICV50876.2021.9388411.
49. F. Hofmann, S. Wurster, E. Ron, and M. Böhmecke-Schwafert, "The immutability concept of blockchains and benefits of early standardization," *Proceedings of the 2017 ITU Kaleidoscope Academic Conference: Challenges for a Data-Driven Society, ITU K 2017*, vol. 2018-January, pp. 1–8, Jun. 2017. https://doi.org/10.23919/ITU-WT.2017.8247004.
50. C. Castellon, S. Roy, P. Kreidl, A. Dutta, and L. Boloni, "Energy efficient Merkle trees for Blockchains," *Proceedings - 2021 IEEE 20th International Conference on Trust, Security and Privacy in Computing and Communications, TrustCom 2021*, pp. 1093–1099, 2021. https://doi.org/10.1109/TRUSTCOM53373.2021.00149.
51. B. R. Williams, "How tokenization and encryption can enable PCI DSS compliance," *Inf. Secur. Tech. Rep.*, vol. 15, no. 4, pp. 160–165, 2010. https://doi.org/10.1016/J.ISTR.2011.02.005.

9 Secure Blockchain-Based E-voting System for Smart Governance

Raja Muthulagu, Pranav M Pawar,
Ashish Kumar Jha, Karan Sharma,
and Kavya Parthasarathy

9.1 INTRODUCTION

One of the pillars of democracy is that elections must be conducted freely and fairly as it serves as a way for the public to express their opinions without fear. In most democracies, the new generation of voters shows reluctance towards the old and outdated process [1]. To overcome this hurdle, E-voting is being promoted as the solution [2, 3]. And as of today, globally prominent organizations have adopted E-voting as the official solution. Estonia used E-voting in 2005, and in 2009 it used E-voting for the national elections and was the first country to do so [4, 5], followed by Switzerland and Norway. Switzerland used E-voting to conduct its state elections [6] and Norway similarly used E-voting to elect its council [7]. However, these electronic methods had their drawbacks. The Estonian and Norwegian systems didn't have open-source codes for casting votes, thus questioning the transparency of the process. The entire system was also susceptible to DDOS attacks, making the elections inaccessible to the voters [8]. As elections hold significant importance in various countries and societies, elections must be dependable, transparent, and free of external tampering. Electronic voting measures have gained lots of momentum, but they must adhere to some baselines, including non-repudiation, integrity, and anonymization of the voter, to make E-voting widely used around the globe [9–11]. One technology that presents foundations based on cryptography that helps achieve solutions to security problems is the concept of blockchain which was proposed by Mr. Satoshi Nakamoto in 2008 [12, 13]. Smart contracts, a variation of blockchain technology, is one way to instil trust in the voters to fight the drawbacks of the Estonian elections [14]. Blockchain, also known as the distributed ledger technology, can be primarily thought of as a database that is decentralized and distributed, which maintains a list of records of data that are constantly growing and further protects the data from tampering and manipulation from unauthorized personnel. It can be thought of as a data structure with a genesis that can share and maintain the different transactions that are being executed. Allowing each user a network connection permits them to add and verify new transactions and hence create blocks. On creation, the block is then broadcasted over the network (Person-to-Person or P2P network) and all the nodes then validate the transaction that takes place. Once validated, the transaction is combined with the already existing blocks to create a new block for the hash/ledger, also called the block's fingerprint. This cryptographic hash is valid until the data in a block is modified. If any alteration takes place to the data in a block, it causes the hash function to change which sends in an indicator that the block has been tampered with. Hence, a secure block is created with every new transaction being added. E-voting using blockchain, thus, provides us with a variety of advantages, viz., greater transparency due to open and distributed ledgers, anonymity from other users, security, and reliability as well as immutability [15].

The applications of blockchain include cryptocurrencies like Ethereum or Bitcoin wherein the transactions of buying, spending, or exchanging cryptocurrencies are being captured and recorded in a blockchain. Not just in cryptocurrencies, blockchain also finds its applications in banks when it comes to processing fiat currencies like dollars and euros making it faster to transfer money. This technology can also be used to keep a record of and transfer ownership of various assets and in the concept of smart contracts also called self-executing contracts which execute when certain conditions are met. Further usage of blockchain includes monitoring of supply chains wherein information such as goods quality and other miscellaneous information is stored onto the blockchain making it easier to keep track and check the supply chain. The last usage of blockchain is the one that we shall be exploring in this project, namely voting systems. To prevent fraud in voting, blockchain is used to make sure that a vote given to a candidate can't be tampered with.

There are notable issues in the case of voting systems and elections in general, for example, when it comes to counting and verifying votes, which is a process that is managed by different equipment and phases. In different counties, the votes can be collected in many ways, including the paper ballot, raising of hands, voting online, and similar methods, but it is noticed that the use of blockchain in order of distributing the election register is an extremely popular approach. In traditional cases wherein votes are recorded, verified, segregated, and counted by a third party/central authority, blockchain decentralizes the controls while the user takes on tasks to be done. As the fundamental concept of blockchain and its use in blockchain helps cut the possibilities of vote tampering, it serves to maintain the integrity of elections. Further in the case of organizations too, it's common to see that since decisions must be made in a swift way alongside the amassing of a lot of feedback and other documents, the pressure can lead to the fall of democracy within an organization and concentration of power in the hands of one. In this case, the concept of cryptographic voting is used which helps ease the decision-making process of different organizations. Such voting can be used to solve the aforementioned problems in different situations relating to involving all necessary stakeholders in important business decisions. This chapter includes the proposed method to create an E-voting portal/website using the technology of blockchain which would have an easily understandable user interface to allow voters to complete the voting process without much trouble while keeping the concept of anonymity, integrity, and security.

The chapter is structured as follows: Section 9.2 talks about the structure, components, and working of a blockchain; Section 9.3 highlights the related work done in the area; Sections 9.4 and 9.5 include the proposed methodology and implementation using smart contracts and consensus algorithm. Section 9.6 contains the results obtained, which is followed by the conclusion in Section 9.7.

9.2 BLOCKCHAIN FOR E-VOTING

9.2.1 BLOCKCHAIN STRUCTURE

The concept of blockchain was first suggested as a research methodology in 1991 which further found its actual application in 2009. In simple terms, blockchain refers to the distributed database that is shared amongst the various nodes/users that are present on a particular network. The mentioned database digitally stores data as required. One of the most common uses of blockchain technology is in cryptocurrencies, where it is employed to maintain a decentralized system that stores transaction records. By doing so, the security and integrity of the transaction data are guaranteed, and the need for a third-party checker is removed, helping to keep transactions secure. A blockchain database differs from a typical database in the way that the data is organized. A blockchain gathers information in the form of clusters or groups, formally known as different blocks, whereas the typical database stores data in the form of tables. These blocks contain information regarding the data/transactions. Each block has a limit when it comes to its storage capacity, and when a particular block is full, it is closed and further linked to other blocks, thereby forming a chain of blocks, i.e., the blockchain. Every new piece of information is first gathered, added to a new block, and then the chain. It must be noted that a timestamp is added to each block at the time when it is added to the chain. Hence, it can be seen that the primary goal of a blockchain is to allow data to be stored and transmitted to various users but make sure it can't be edited, thereby providing a foundation base for immutable ledger transaction records that can't be deleted or modified. Therefore, a blockchain is also called a distributed ledger technology. In the years since its inception, blockchain has found various uses and applications due to the creation of different cryptocurrencies, smart contracts, non-fungible tokens, and decentralized finance applications.

9.2.1.1 Working on a Blockchain

When considering the formation of the blockchain and how it is created. Any user that is connected to the internet and is a part of the network can place a transaction into the blockchain. In case a new transaction is to be added, the user first enters it. This newly added transaction is then broadcasted over the blockchain peer-to-peer network containing computer nodes run by users present all over the globe. These nodes try to validate this new transaction added by solving certain problems or equations. Once this newly added transaction is validated, it is considered legitimate and added to a block. All other transactions added also go through the same process and are ultimately added to the same block unless it is full. If the block is full, the newly verified transaction would be placed within a new block. These blocks of data records are then connected to form a chain. Hence, the transaction is completed. Figure 9.1 explains the blockchain creation process.

9.2.2 ADVANTAGES OF USING BLOCKCHAIN IN ELECTRONIC VOTING

9.2.2.1 Accuracy

All transactions added onto the blockchain are validated by hundreds of nodes present over the blockchain network, thereby removing the chances of there being any kind of human error in the process of verification resulting in greater accuracy. In case a computer node makes a computational error, it would be made only to a single copy of the

Secure Blockchain-Based E-voting System for Smart Governance

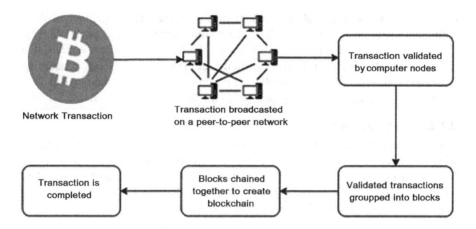

FIGURE 9.1 Blockchain creation wherein a new transaction is added to the blockchain.

blockchain. If such an error would have to spread over the entire network, approximately fifty-one per cent of the nodes present would have to make the same computational error which is highly unlikely when the size of the network is continuously growing. This makes sure that there would not be any errors when it comes to validating the votes and there would not be any vote tampering in the case of electronic voting.

9.2.2.2 Reduction of Cost

When a transaction needs to be confirmed, a customer often pays the bank; this price is referred to as the transaction fee. Such a fee is present in other activities, for example, getting a document signed by a notary. As the concept of blockchain allows other nodes to verify the transaction, the need for any kind of third-party verification is eliminated, and hence the associated cost is also removed. In the case of electronic voting, there need not be people employed to count and tally the votes, thereby reducing associated costs and effort.

9.2.2.3 Security

Any of the information present in a blockchain is not stored at a central location, it is rather copied and present across a network of nodes. Every node on the network updates its blockchain as a new block is added to the chain in order to reflect the modification and the new block's insertion. As this information is not stored in a central database and is spread across the network, this makes blockchain more secure. In case a copy of the blockchain is broken into by a hacker, only that copy would be compromised but the entire network would still be safe, thereby strengthening the security. Regarding an electronic voting system, the likelihood of vote tampering is significantly decreased thanks to the security measures offered by a blockchain.

9.2.2.4 Efficiency

When a transaction is placed, in cases where there is a central approving authority, it is noticed that it may take a few days to settle the transaction. An example of this is, if one tries to deposit a check on a Friday, the funds may not be seen in the required account until a few days later, said Monday. Unlike financial institutions, blockchain operates 24/7, meaning that all transactions can be completed in a minimum of about ten minutes and this transaction will be considered secure within a couple of hours. This aspect of blockchain is useful for trades that take place across borders that normally would take more time due to differences in time zones. Hence, concerning electronic voting, people all around the world would be able to cast their vote and see it being cast instantaneously, i.e., there would not be any kind of delay.

9.2.2.5 Transparency

Most blockchains have their code available for everyone to see, i.e., they are open-source software. This allows auditors to review the codes to determine how secure it is. As there is no authority controlling the code, it can be edited or improved by anyone. If the majority of users of the software agree that the upgradation of the code is safe and worth it, then the software would be updated.

9.2.3 COMPONENTS OF THE E-VOTING SYSTEM

A blockchain tries to solve many real-world problems. Based on the requirement, different consensus algorithms or smart contracts can be used to build an E-voting system. Certain choices need to be made regarding the type of blockchain network to be used, the consensus algorithm to be implemented, integrations to be made, and what level of anonymity is to be maintained.

9.2.3.1 Blockchain Network

There are primarily three different types of blockchain network architectures: Category 1 includes public networks with permissionless access, such as the Ethereum and Bitcoin networks; Category 2 includes public networks with permissioned access, such as ripple; and Category 3 includes private networks with permissioned access, such as FiberChain and Bankchain. Figure 9.2 shows these categories.

The choice of network architecture depends on the level of decentralization, the amount of information to be displayed to the public, and the fee per transaction. When partial decentralization is required, networks that are permissioned are preferable. If maximum decentralization is required, a public, a permissioned blockchain is better. With public networks, all information pertaining to a transaction is accessible, allowing for real-time election monitoring. However, if specific information needs to be kept private from the public, a private blockchain network would be more suitable. All public blockchain networks will require a transaction fee that acts as a security measure to prevent denial of service (DoS) attacks.

Public networks with permissionless access

Operate with untrusted members

Public networks with premissionless access

Operated only with trusted members

Private networks with permissioned access

Operate with trusted members of a defined community

FIGURE 9.2 Blockchain network architectures.

9.2.3.2 Consensus Algorithm

A consensus algorithm refers to a mechanism that maintains coordination between users or machines by making sure that all the agents present inside the system can agree on certain truth points in case certain agents fail, i.e., creating a fault-tolerant system. Consensus algorithms are divided into two categories: competitive and non-competitive. Competitive algorithms include proof of work, proof of stake, and delegated proof of stake. Non-competitive algorithms include the Paxos algorithm, Raft algorithm, and Practical Byzantine fault tolerant (PBFT) algorithm.

9.2.3.3 Anonymity

In the default setting, blockchain networks are public in nature, causing users to get access to details of a transaction. But it must be noted that in the case of online voting, it must be made sure that the ballot made by a vote is kept secret. Hence, a vote should not be linkable to the voter who cast it; hence, there must be anonymity of the vote given by the user. In certain cases, including voting in parliament and stakeholder casting votes, it need not be necessary to keep the ballot a secret.

9.3 RELATED WORK

Substantial work has been done in the field of electronic voting using blockchain, some related works are as follows. Francesco et al. [16] build the voting system 'crypto-voting system' using a two-linked blockchain, one-way pegged sidechains that act as an extension to the blockchain, and Shamir's secret sharing method. This method further enables the creation of new characteristics by avoiding writing inside the primary blockchain. Recording of eligible voters alongside voting operation is stored in the first sidechain and the vote count is recorded in the second. Khan et al. [17] use the multichain platform to create an efficient scheme for E-voting while utilizing the concepts of transparency and cryptography to achieve an effective solution. In order to determine which voting systems satisfy which criteria, such as reliability, anonymity, security, etc., Garg et al. [18] conducted a review of the literature on various techniques and solutions for the problems with electronic voting and provided a comparative study. However, they came to the conclusion that the majority of these systems needed better authentication techniques to confirm user identity. Kshetri and Voas [19] try to evaluate the pros and cons of the BEV standing for Blockchain-enabled voting and list the advantages and challenges faced by this methodology. It was noted that this methodology provided trust, accuracy, and security, and the main issue faced is that accessibility to the public is inhibited due to its complexity and fluctuations in a broadband connection. Yavuz et al. [20] have implemented an E-voting system using smart contracts on the Ethereum platform to keep the system secure and transparent and reviewed the time taken for the vote given by a voter to get cast and therefore have judged the efficiency of the system, but this methodology is limited to small-scale elections. Khan et al. [21] attempt to figure out the scalability constraint for an electronic voting system by determining the effect of several parameters including block size, speed of vote consideration, and population, and checking their effect on the efficiency of the system. A scenario in which all parameters are at their best and therefore have the highest efficiency is unlikely; it is seen when the experiments are conducted by varying these various parameters that different tweaking results in a high performance of one of the parameters with a trade-off of the others. Widayanti et al. [22] make use of three separate models – integrate model, voting model, and withdrawal model – to build an overall E-voting system that uses the concept of elliptical curve cryptography (ECC). The designing of the system has been done to prevent vote forgery and meet other parameters. The main issue is that the ECC technique used is susceptible to attacks made from a quantum computer. Chaudhari [23] uses the concepts of smart contracts as used in Ref. [20] but focuses on maintaining privacy more. The methodology included the creation of smart contracts, authentication of the user, and verifying their identity, applying both PoW and proof of voting (PoV) algorithms followed by dealing with faulty nodes. It checks for the chance of failure and the efficiency of the smart contracts, and it was noticed that the PoW methodology reported a higher chance of failure when compared to the PoV method. Yousif et al. [24] provide a comparative study between the need for privacy and security in an E-voting system by comparing the various methods used based on certain parameters, including eligibility, anonymity, reliability, transparency, fairness, robustness, etc..

Also, it must be noticed that privacy and security are both essential in an E-voting system and based on different methods, satisfying these two needs have been improved. Also, it must be noted that more work is needed on the subject of verification of the authenticity of the votes that have been cast alongside anonymity, speed of casting of votes, and their accessibility. Hjálmarsson et al. [25] test an electronic voting system using Ethereum and Solidity algorithm to use smart contracts and increase security and decrease the cost of building using HTML, CSS, and JavaScript alongside NodeJS, and it is noticed that transparency, security, and audibility are satisfied well. Thuy et al. [26] provide another proposal for an E-voting system using blockchain that minimizes the concept of centralization and increases the concept of fairness. The process involves the creation of the interface to be used, followed by the creation of an application server and regulator for the ballot alongside the Ethereum node and the ballot contract; it is further noticed that the model helps decrease the cost and improves the experience of the user. Vivek et al. [27] create another decentralized voting system based on blockchain on the multichain platform which resolves the issue of the sawtooth framework which ensures scalability of the system, fairness when it comes to voting, and user anonymity on a small scale. Jafar et al. [28] create a multichain platform-based voting system, focusing on improving security as mentioned in Ref. [27] alongside confidentiality and integrity of the vote. It is implemented by creating the backend alongside the nodes, specifying elements like storage capacity, verification of the identity of the voter, and lastly the casting of the vote. Haibo [29] uses the concept of ECC to secure an E-voting system that works through the process of generating the signature of the voting block using a private key and its verification using the public key. Khan et al. [30] finally check the possibility of the transaction malleability attack within an application based on the concept of blockchains. In order to come to the conclusion that attack chances are prominent when the time between the sent transaction that is honest and the malleable transaction is very short, the paper attempts to explain circumstances that could potentially lead to that attack and checks various conditions to see if there is any prominent side effect if the attack is carried out on the application..

Ref. No	Objective	Problem Statement	Methodology	Advantages	Disadvantages	Software Used	Performance Measure
[16]	Defining a new voting system; crypto-voting	Integrating management procedures and events of elections	Shamir's secret sharing approach	Provides security, remote voting	The actual code has not been written and tested for the same	Blockchain and sidechain	Efficiency
[17]	achieve an effective scheme for E-voting	Using multichain to create an effective E-voting system	Using blockchain to implement an E-voting system	Maintains privacy, and checks eligibility, convenience, and verifiability	Needs to improve the resistance to blockchain technology	Java EE within NetBeans using Glassfish server	Effectiveness
[18]	Finding a method to solve different voting issues	Providing a view of E-voting thematically	literature review	NA	Biometric authentication of user required	NA	Checking the presence of all attributes in the voting system
[19]	Reviewing the pros and cons of blockchain-enabled voting (BEV)	Listing the advantages, and challenges of implementing BEV	Literature review and research	Providing security, accuracy, trust	Public accessibility is inhibited due to complexity, limited broadband access	BEV	NA
[20]	Testing E-voting application as a smart contract	Using blockchain to provide a secure voting environment	Use of the Ethereum platform and blockchain to build a voting system	Creation of a transparent platform for voting	Scope limited to small-scale elections	Ethereum wallets and Android platform	Time is taken for a vote to get cast
[21]	Finding scalability constraints for an E-voting system	Determining the effect of population, block size, speed, etc. on the efficiency of the E-voting system	Experiments using blockchain in different conditions	In different cases, different performance metrics give satisfactory results	Trade-off present between parameters, hence best in all cases not likely	Multichain platform	Efficiency and scalability of the model, security, performance

(Continued)

Ref. No	Objective	Problem Statement	Methodology	Advantages	Disadvantages	Software Used	Performance Measure
[22]	Design and integrate voting, credential, and withdrawal model	To use blockchain techniques and secure E-voting platforms	Designing different components to avoid vote forgery, authenticate, and meet other requirements	Practically an E-voting system can be implemented using blockchain to solve the vote forgery issue	ECC public cryptography used is vulnerable to quantum computer attacks	Linux platform, Python language	Efficiency to prevent vote forgery
[23]	Designing an E-voting system using a blockchain and proof of voting (PoV) algorithm	Using smart contracts to secure E-voting system while maintaining user privacy	Creating smart contracts, authenticating, applying PoV, and dealing with faulty nodes	Security increased, costs lowered, and less power was consumed using the PoV model	The proof of work model gives relatively more chance of failure compared to PoV and the smart contracts are slow	Blockchain technology	Based on the chance of failure and smart contract efficiency
[24]	Providing a study comparing about security and privacy needs of E-voting system	To compare various methods taken to build E-voting systems in blockchain and compare them	Literature review of various methods used and comparing them on certain parameters	Amongst all papers, it is noticed that privacy and security have been improved	More work is needed when it comes to verifying the authenticity of votes, anonymity, scalability, speed, accessibility	NA	Eligibility, anonymity, fairness, auditability, consistency, robustness, security, vote privacy
[25]	Testing the E-voting system as a smart contract by using Ethereum and Solidity algorithm	Introducing an E-voting system using blockchain and smart contracts to increase security and lower the cost	Using HTML, CSS, and JavaScript as front end alongside NodeJS, and Blockchain in the backend	The use of blockchain provides transparency and security, makes it auditable and more open	NA	HTML, Bootstrap Blockchain, Ganache-CLI.	Security aspect and number of transactions added per second

(*Continued*)

Secure Blockchain-Based E-voting System for Smart Governance

Ref. No	Objective	Problem Statement	Methodology	Advantages	Disadvantages	Software Used	Performance Measure
[26]	Proposing an E-voting system utilizing blockchain, i.e., Vitreum	Creating an E-voting system that minimizes the use of central authority and increases fairness	Creating user interface, application server, online ballot regulator, Ethereum node, ballot contract	The proposed model helps reduce cost, improves user experience, is secure and accurate	On Rinkeby, errors are faced when lots of ballots are added to the network in a brief time	Blockchain, Ethereum platform, NodeJS, RabbitMQ	Privacy, accuracy, security, effectiveness, fairness, verifiability, etc.
[27]	Creating a secure, fair, and decentralized system for E-voting	Use blockchain and create an E-voting system, resolve hyper-ledger saw-tooth framework	Creating various entities needed and then applying security mechanisms	Ensures anonymity of users, scalability, and fairness in voting	Further exploration is required to implement on a large scale	Blockchain, multichain platform	Anonymity and security
[28]	Creating a multichain-based E-voting system	Building a secure E-voting system and showing that auditing ensures the system is authentic	Creating backend and nodes, specifying storage, using trusted third party (TTP) to verify voter validity, casting vote	Is secure and scalable in approach, limits each user to cast just one vote	Delays encountered as each voter's private message is to be verified by a TTP	Multichain, blockchain platform	Integrity, anonymity, data confidentiality, and reliability
[29]	Using elliptical curve cryptography (ECC) in the voting system	Securing an E-voting system using ECC	Generating vote block signature through private key and verifying using public key	There is an elaborate verification and evaluation procedure for voting blocks	PKI database which is used is vulnerable	Blockchain and ECC	Security, validity, fairness
[30]	Checking whether transaction malleability attack is possible within blockchain application	Highlighting circumstances that can lead to a transaction malleability attack	Using the multichain platform to simulate the attack and checking different conditions in which the attack is prominent	The chance of the attack getting into the block is less when compared to receiving a transaction that's honest in the block	Chances of attack are more prominent when the time between sending of honest transaction and transaction (malleable) is less	Multichain blockchain platform	Network delay, block generation rate

(*Continued*)

Ref. No	Objective	Problem Statement	Methodology	Advantages	Disadvantages	Software Used	Performance Measure
[31]	To perform a literature review systematically on the scalable blockchain-based voting technologies present	Identification of performance parameters and avenues for improvement in current technologies	Performed SLR on popular platforms like IEEE Xplore Digital Library, SpringerLink, Scopus, ACM Digital Library, and ScienceDirect	Provides parameters to compare various present technologies and contrasts with the pre-existing electronic voting system	Since the present day holds vast avenues for the development of E-voting technologies, comparing all of them gets difficult, and finding the correct metrics for comparison can be challenging	Multichain, blockchain platform	The popularity of various tools and blockchain technologies in the current scenario
[28]	To identify the benefits of blockchain-based electronic voting systems to counter the fallacies in the current system	To review and challenge the current electronic voting systems against the blockchain-based E-voting systems	Performed SLR on the existing E-voting technologies as well as the popular blockchain-based E-voting systems	The study concludes the benefits and advantages of using blockchain to modernize, enhance security and provide transparency to the voting process	The blockchain-based E-voting system needs a robust security feature to prevent malpractitioners from tampering with the voting process to conduct a fair election process	Multichain, blockchain platform along with the present E-voting methodologies	Used parameters like transactional privacy, energy efficiency, and acceptableness to compare the technologies in hand
[32]	To maintain the privacy of a voter in the election process using blockchain-based E-voting systems	Encrypting the votes when being submitted to the Blockchain in a secured fashion to preserve the voter identity	Divided the entire voting process into three different steps. These steps include the setting up of the system, followed by the voting process, and finally the tallying step	Preserves the voter's identity and provides additional security to the privacy of the voter by encrypting the votes cast before storing them in the blockchain	Designed and tested for small to medium-scale election scenarios	Used a novel register algorithm to preserve voter identity	Comparing the results against a set of experiments conducted comprehensively designed to identify the workload, transaction cost, and bound parameters

We have proposed and implemented two kinds of E-voting systems. One is based on smart contracts written in the Solidity language using the Ethereum network and integrated with a graphical user interface (GUI), whereas the other is run on a command prompt based on parts of the PBFT algorithm and the Ripple Protocol Consensus (RPC) algorithm. We shall be discussing both approaches one by one alongside the components of each proposed solution.

9.4 PROPOSED E-VOTING SYSTEM BASED ON SMART CONTRACTS

A long-due challenge in the case of electronic voting has been building a system that secures the privacy of the voter alongside keeping elections fair and free of tampering. This system is made using the Ethereum network and is based on the concept of smart contracts. A smart contract can be defined as a contract that is self-imposed and present within a computer code that is managed using blockchain. The code comprises regulations that set forth how the parties will communicate with one another and make decisions regarding the terms of the contract. When the defined rules have been met, the contract will be enforced automatically. A smart contract provides you with a framework to maintain efficient control between different parties regarding access rights and assets that are tokenized. It can be thought of as a special kind of program that runs on a virtual machine which is further made into a distributed ledger. The creation of a smart contract is rather simple: developers define their requirements for what a smart contract must do when faced with a particular event. Examples of events include payments to be authorized or a received shipment. These occasions may also be more complicated, such as when processing a trade. The developers then develop the logic and test it to check if it works as required. After the development phase, the security aspects are taken care of, and once approved, the contract is deployed. Figure 9.3 shows the principle of working behind a smart contract.

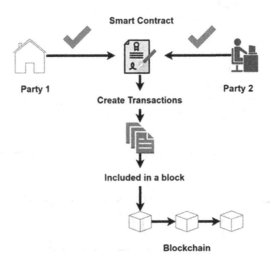

FIGURE 9.3 Blockchain and smart contract.

The process of self-verification of the contract is done using the concepts of data interpretation wherein each node present on the network will confirm and guarantee the proper execution of every contract. This spares the developers from having to keep tabs on how the contract is being carried out. When the conditions meet the requirement of the contract, the contract will run automatically; this allows the legal obligations between parties to be mapped into the code of the contract itself. A trigger, such as an expiration date, may be used to activate the contract while it is in use. In this chapter, smart contracts have been implemented in an E-voting system.

9.4.1 Concept behind Voting System Using Smart Contracts

The tools used include Ganache, truffle, a node package manager (NPM), and Metamask. The Ganache will act as the actual internal blockchain which shall be accessed using Metamask whereas truffle handles the task of importing the smart contracts onto the blockchain as shown in Figure 9.4. Using Ethereum's cryptocurrency called Ethers (ETH) is necessary to give a user a wallet address corresponding to a particular user's account. A fee known as gas must be paid when the user wants to write a transaction into the blockchain. When the votes are cast, the voting process is completed by certain nodes in the network called miners who compete to complete the transaction and get awarded certain Ether (which was originally paid by users when they had to vote). This is a general idea but, in our case, instead of using nodes to complete the voting, we use Ganache for mining.

9.4.2 Software and Tool Requirements

To build an E-voting system using the concepts of smart contracts, the following dependencies are required:

- Node package manager (NPM)
- Truffle
- Ganache blockchain
- Metamask cryptocurrency
- Required language to code: Solidity, JavaScript, HyperText Markup Language (HTML), Cascading Style Sheets (CSS)

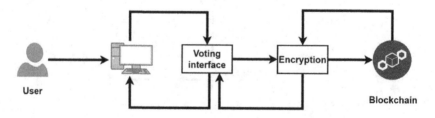

FIGURE 9.4 Proposed voting system using blockchain and smart contracts.

9.4.2.1 Node Package Manager

As the name suggests, NPM, which is a command line tool, is a package manager that installs, updates, manages, and uninstalls the different NodeJS packages present in the application. The tool has two modes of operation globally. In the local mode of operation, only a specific directory of the application will be affected, whereas, in the global mode of operation, all the NodeJS packages in the application will get affected.

9.4.2.2 Truffle

The truffle framework is an immensely powerful tool that works with Ethereum-based smart contracts. The framework is used for compilation purposes alongside the actual deployment and linking of the contracts. Further, the tool also provides a platform to test contracts that are automated amongst other uses.

9.4.2.3 Ganache Blockchain

Formerly known as Testrpc, Ganache is a blockchain platform used for the development of distributed applications for Ethereum and Corda. It is used over the cycle of development, allowing one to develop, manage, deploy, and further test the working of your applications in a safe space. A free account in Ganache gives you 10 free standard addresses in Ethereum alongside the private keys for each of them. By using the Ganache, the process of mining gets eliminated and the transaction gets automatically confirmed. Operable on Windows, Linux, and macOS, it is usable in UI form alongside command line form.

9.4.2.4 Metamask

Metamask is an open-source software that has a GUI, providing you with a wallet with cryptocurrency which is used to interact with the Ethereum blockchain. Users are allowed to access their Ethereum wallet through the software which can thereby be used to interact with the application being designed. In simple terms, Metamask acts as a bridge between the blockchain, and the browser being used.

9.4.2.5 Solidity

Solidity refers to a high-level programming language following the concepts of object-oriented programming which is used to write smart contracts. It is used to implement smart contracts for different platforms regarding blockchain mainly Ethereum. Solidity contains the structural syntax and operators like JavaScript containing type values of integer, Boolean, address, and string. Further, it acts as a methodology to generate a machine-level code and convert it into simple understandable instructions.

9.5 PROPOSED E-VOTING SYSTEM BASED ON CONSENSUS ALGORITHMS

A consensus algorithm is a mechanism that is fault-tolerant and is used in blockchain to gain agreement on the state of the network. The consensus algorithm used here takes inspiration from the PBFT algorithm and the RPC algorithm. The PBFT algorithm was designed to correct several issues present in the solutions regarding byzantine fault tolerance (BFT). The BFT allows nodes in a distributed network to reach consensus (agreement on a value) even if some nodes in the network do not respond or give a wrong response. Its objective is to protect the network against system failures by using collective decision-making methods, which means that choices are made by both the right and wrong nodes, reducing the influence that the defective nodes have.. BFT can thereby be achieved if the nodes that are currently working can reach a consensus on their values within a network. Regarding messages that are not present, a default value can be given, i.e., it is assumed that the message received from the node is 'wrong,' if it doesn't get received within a specific time frame. Antagonistically, if most of the nodes end up responding with the right value, then a default response in such a case can be assigned. It was proved by Leslie Lamport that if $3m+1$ processors work properly, it is possible to attain consensus only if there are only 'm' processors at most that are faulty which simply means that consensus can only be reached if two-thirds of the processors are working properly. The PBFT tries to give a byzantine state machine that is practical and can work when there are faulty nodes present in the network system. In a PFBT-enabled system, which functions only when the maximum number of faulty nodes is less than or equal to one-third of the total nodes present; nodes are ordered in a sequential manner wherein one node is considered the leader and others are referred to as backup nodes. Any node in the system can be a primary node, and in case this node fails, then a secondary/backup node can transition to form a primary node. The main objective here is to make sure that all honest nodes can reach a form of consensus through the use of the majority rule. The round for consensus has four basic phases:

- First, a request is sent to the lead node from the client.
- The request is then broadcasted by the lead node to all the backup nodes.
- All nodes then execute the necessary task and send a response to the client.
- The request is successfully served when the client receives $m+1$ replies from the nodes containing the same answer.

The RPC algorithm refers to a process that is carried out by the nodes present in the blockchain network every few seconds to reach a consensus. After all the nodes successfully agree, the ledger remains closed. The steps involved in the algorithm are as follows:

- All known transactions that are valid and not part of the ledger are collected by each server and made public. These unconfirmed lists of transactions are otherwise called candidate transactions.
- All candidate lists are collected by each server and transactions that get the minimum requirement of positive votes are proceeded with.
- Lastly, it's made sure that each transaction has secured a minimum of eighty per cent of a server's votes, and such transactions are added to the ledger.

The consensus algorithm used in the E-voting system is based on parts of the PBFT algorithm and the RPC algorithm. It states that:

- For all the nodes present in the network, broadcast the hash value of the previous block present in the chain.

- Then broadcast the candidate set of transactions and validate the global candidate transaction set.
- Lastly, broadcast the approvals for the transactions and thereby aggregate all the transactions that are approved into a new block.

9.6 IMPLEMENTATION AND RESULTS

9.6.1 Vote Casting

9.6.1.1 Working

The working of voting system is illustrated in Figure 9.5. To vote using the web application, the voter must first log onto the website, and use the Metamask extension to connect with the Ganache blockchain which is the local blockchain network. The constructed smart contract is moved onto the established local blockchain network using the truffle framework. Once the connection takes place between Ganache and Metamask, the website running on the local host will automatically pop up and the

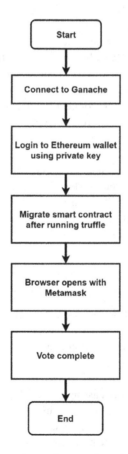

FIGURE 9.5 Working of voting system using smart contracts.

user will be able to see the candidates for whom they can vote. On Ganache, when one logs in, they get identified with 10 free accounts with their account address, and private and public keys. After using the private key to connect to the Ethereum wallet on Metamask, the user can make their vote. Once the user decides whom to vote for and clicks on the vote, the vote is placed for the user and the required transaction fee, i.e., a small amount of Ethers is deducted. It is observed that if the same account even attempts to vote again, the voting option isn't available, and the popup comes up that the user had already voted and hence cannot vote again, i.e., the transaction will fail in that case. One will see that after casting a vote, the user's vote appears in the total number of votes cast for the candidate they choose. As a result, the entire number of votes is displayed in real time, giving users the opportunity to verify and confirm their findings. Once the user finishes voting, it will be observed that the Ethereum wallet for that account will not have the same amount of 100 ETH that was present initially; hence, if the user attempts to vote again, the transaction won't be allowed, and the transaction of voting will fail. As mentioned earlier, the process of mining will not take place through the nodes present in the network; Ganache is given the power to perform the mining operations on behalf of the nodes.

9.6.1.2 Implementation

Initially, the local blockchain system is set up on Ganache. In order to deploy the smart contract onto the local Ganache blockchain, NPM is then run concurrently with truffle migrate. It is noted that for the transaction to be made, a certain amount of transaction fee would be deducted.

On the side, Metamask is run to connect the account using its private key to the Ethereum wallet to deduct the required cryptocurrency for that user's vote. Then when Metamask is connected, the browser opens with the options to vote. The user is not offered the option to vote again after choosing the candidate they wish to support and casting their vote because it can be observed that the vote has already been cast. As shown in Figure 9.6, we can see the decentralized voting network election results as the vote has been considered for the user.

After the vote is cast, the transaction fee is deducted for the account, implying that the vote has been cast successfully.

9.6.2 After Vote Casting

9.6.2.1 Working

The code has been run on the command prompt. The code has been made in a way that it can simulate an election based on certain cases alongside having the normal (interactive) mode of how an election is normally created. The simulation cases include the following:

1. A voter who is valid and is casting a vote that is also considered valid in which case the network simulates consensus taking place between fifty nodes and displays the election results.
2. An unknown voter who is trying to cast a vote in which the simulation log shows an error saying that the unknown vote is not present in the voting roll list.

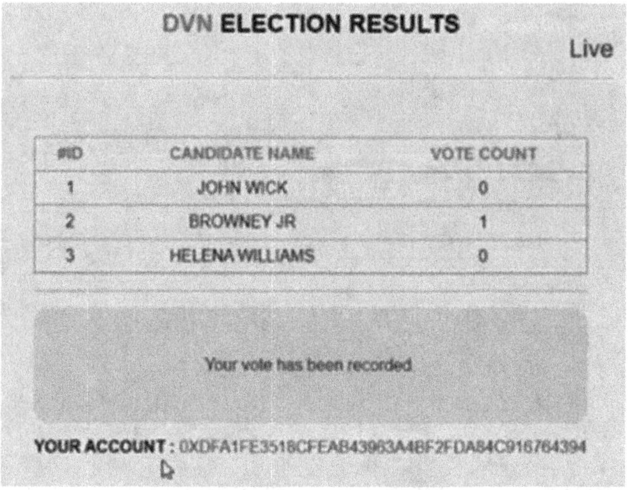

Secure Blockchain-Based E-voting System for Smart Governance 179

DVN ELECTION RESULTS

Live

#ID	CANDIDATE NAME	VOTE COUNT
1	JOHN WICK	0
2	BROWNEY JR	1
3	HELENA WILLIAMS	0

Your vote has been recorded

YOUR ACCOUNT: 0XDFA1FE3518CFEAB43963A4BF2FDA84C916764394

FIGURE 9.6 The user has cast the vote.

3. A voter who is considered valid but tries to cast more than one vote in which case the simulation throws an error saying that the voter does not have enough claim tickets to be able to cast another vote. The claim ticket value alongside the voter list is present in the JSON file wherein the name of each voter is present alongside the number of claim tickets owned which is one in this case per voter. This is hardcoded by the developers and is not available to the public to change.

The system demonstrates the consensus process taking place between the nodes present in the blockchain and is also able to simulate the behaviour of faulty nodes, thereby demonstrating the power that consensus holds. All the transactions made are even stored in the form of a log. When the voting is opted to take place in the interactive mode which is normally how an election would take place. The user has displayed a list of menu items allowing them to vote, look up their voter ID which is uniquely assigned to each voter to make sure that even if two people have the same name, they are still allowed to cast the votes, check for current results statistics, an option to view the logs of the transaction votes and the last option being to exit and finish the election, thereby displaying the results.

When a registered voter tries to vote, they are authenticated by asking them to type their full name into the system. When they do, the nodes verify the related check in the backend, and the corresponding voter ID is then presented. The user is then asked to cast their vote for the respective candidate by voting based on their ID number and once the choices are made, they can validate their choice or invalidate and change their response until the voter is satisfied. Once the voter casts their vote, in the backend, the fifty nodes try to attain consensus and thereby approve the transaction. And when this is approved, the voter has successfully cast their vote and can see their vote being counted for their candidate in real time.

```json
[
    {
        "name": "Temp_Voter",
        "num_claim_tickets": 1
    },
    {
        "name": "Rahul CS",
        "num_claim_tickets": 1
    },
    {
        "name": "Raja Muthalagu",
        "num_claim_tickets": 1
    },
    {
        "name": "Pranav Pawar",
        "num_claim_tickets": 1
    },
```

FIGURE 9.7 Voter list.

9.6.2.2 Implementation

Firstly, to see the voters that have been registered, Figure 9.7 shows the JSON file containing the voter list alongside their ID and claim ticket to vote.

On running the code, the initial page is displayed to the user in the normal/interactive mode.

Consider the user who chose a contestant as his/her vote, to verify if their ID has been registered to vote. In that case, the voter is asked to enter their name and their corresponding ID is displayed if valid. Once that is executed, the consensus is reached, and the transaction is considered valid. Then the menu shows up again for further voting.

When the voter makes a choice, they are asked to authenticate themselves by signing their names, and their corresponding claim tickets are retrieved if valid and they are allowed to vote. The voter is then asked to cast their vote for president and vice president, thereby validating the same. Once consensus is made, the votes are counted, and the next consensus round begins alongside displaying the menu again. It must be noted that once the claim ticket is executed fully and the vote is cast, the record is set, and the value set initially becomes zero and hence the voter can't vote again.

Consider that the user finally chooses to check the status of the result. It is seen that their vote has successfully been placed and the menu is ready for the next user to cast their vote.

When the menu appears after every voter has completed their vote, the option to leave while ending the elections becomes available, and the results are then made public for every post.

9.7 CONCLUSIONS

Election tampering is a typical occurrence in today's elections, and because elections are a crucial component of a nation's advancement, it is crucial to ensure that they are handled fairly and securely. With the help of blockchain-related technologies such as smart contracts and a consensus algorithm, this chapter has proposed two methods for holding elections online. The E-voting system is built using smart contracts, as it is integrated with a GUI, when hosted It can be accessed by anyone all over the world to vote, solving one of the main problems when it comes to voting. It must also be noted that as the voter shall be using their private key to log in and cast the vote, it makes the platform secure and tamper-free, as the user will only be able to go ahead with the transaction once. The second algorithm supplied has been designed to work in an offline setting wherein voters can come to the polling booth nearest to their location and can cast their vote. It can be made such that anyone all over the world can access it, but it has its own set of risks when it comes to keeping the servers and other endpoints. As part of our ongoing work, we will connect the voting system to more advanced verification methods, such as connecting it to databases that store Social Security numbers (in the case of the United States) and Aadhar card numbers (in the case of India).

REFERENCES

1. L. C. Schaupp, and L. Carter, "E-voting: From apathy to adoption," *J. Enterp. Inf. Manag.*, vol. 18, no. 5, pp. 586–601, 2005.
2. W. D. Eggers, *"Government 2.0: Using Technology to Improve Education, Cut Red Tape, Reduce Gridlock, and Enhance Democracy,"* Rowman & Littlefield, 2007.
3. T. M. Harrison, T. A. Pardo, and M. Cook, "Creating open government ecosystems: A research and development agenda," *Future Internet*, vol. 4, no. 4, pp. 900–928, 2012.
4. Ü. Madise, and T. Martens, "E-voting in Estonia 2005. The First Practice of Country-Wide Binding Internet Voting in the World," *Electronic voting, 2nd International Workshop*, Bregenz, Austria, August 2–4, 2006.
5. M. A. Uddin, A. Stranieri, I. Gondal, and V. Balasubramanian, "A survey on the adoption of blockchain in IoT: Challenges and solutions," *Blockchain: Res. Appl.*, vol. 2, no. 2, pp. 1–49, June 2021.
6. J. Gerlach, and U. Grasser, *"Three Case Studies from Switzerland: E-Voting,"* Berkman Center Research Publication, 2009.
7. I. S. G. Stenerud, and C. Bull, "When reality comes knocking Norwegian experiences with verifiable electronic voting," *Electronic Voting*, vol. 205, pp. 21–33, 2012.
8. S. T. Alvi, M. N. Uddin, L. Islam, and S. Ahamed, "DVTChain: A blockchain-based decentralized mechanism to ensure the security of digital voting system voting system," *J. King Saud Univ. Comput. Inf. Sci.*, vol. 34, no. 9, pp. 6855–6871, 2022.
9. K.-H. Wang, S. K. Mondal, K. Chan, and X. Xie, "A review of contemporary e-voting: Requirements, technology, systems and usability," *Data Sci. Pattern Recognit.*, vol. 1, no. 1, pp. 31–47, 2017.
10. V. Agate, A. De Paola, P. Ferraro, G. Lo Re, and M. Morana, "SecureBallot: A secure open source e-voting system," *J. Netw. Comput. Appl.*, vol. 191, p. 103165, 2021.
11. R. Anane, R. Freeland, and G. Theodoropoulos, "e-Voting Requirements and Implementation," *The 9th IEEE International Conference on E-Commerce Technology and The 4th IEEE International Conference on Enterprise Computing, E-Commerce and E-Services (CEC-EEE 2007)*, Tokyo, Japan, 2007, pp. 382–392.

12. C. A. Ul Hassan, M. Hammad, J. Iqbal, S. Hussain, S. S. Ullah, M. Mosleh, and M. Arif, "A liquid democracy enabled blockchain-based electronic voting system," *Sci. Program.*, vol. 2022, Article ID 1383007, 2022.
13. C. K. Adiputra, R. Hjort, and H. Sato, "A Proposal of Blockchain-Based Electronic Voting System," *2018 Second World Conference on Smart Trends in Systems, Security and Sustainability (WorldS4)*, London, UK, 2018, pp. 22–27.
14. "What Are Smart Contracts? A Beginner's Guide to Smart Contracts", *Blockgeeks*, 2016, Available at: https://blockgeeks.com/guides/smartcontracts/
15. F. Sheer Hardwick, A. Gioulis, R. Naeem Akram, and K. Markantonakis, "E-Voting With Blockchain: An E-Voting Protocol with Decentralisation and Voter Privacy," *2018 IEEE International Conference on Internet of Things (iThings) and IEEE Green Computing and Communications (GreenCom) and IEEE Cyber, Physical and Social Computing (CPSCom) and IEEE Smart Data (SmartData)*, 2018, pp. 1561–1567. https://doi.org/10.1109/Cybermatics_2018.2018.00262.
16. F. Francesco, L. Maria, P. Filippo, and P. Andrea, "Crypto-voting, a Blockchain-based e-Voting System, *"In Proceedings of the 10th International Joint Conference on Knowledge Discovery, Knowledge Engineering and Knowledge Management (IC3K 2018)*, 2018, vol. 3, pp. 223–227.
17. K. M. Khan, J. Arshad, and M. M. Khan, "Secure digital voting system based on blockchain technology," *Int. J. Electron. Gov. Res.*, vol. 14, pp. 53–62, 2018.
18. K. Garg, P. Saraswat, S. Bisht, S. K. Aggarwal, S. K. Kothari, and S. Gupta, "A Comparative Analysis on E-Voting System Using Blockchain," *2019 4th International Conference on Internet of Things: Smart Innovation and Usages (IoT-SIU)*, 2019, pp. 1–4. https://doi.org/10.1109/IoT-SIU.2019.8777471.
19. N. Kshetri, and J. Voas, "Blockchain-enabled e-voting," *IEEE Softw.*, vol. 35, no. 4, pp. 95–99, July/August 2018. https://doi.org/10.1109/MS.2018.2801546.
20. E. Yavuz, A. K. Koç, U. C. Çabuk, and G. Dalkılıç, "Towards Secure e-Voting Using Ethereum Blockchain," *2018 6th International Symposium on Digital Forensic and Security (ISDFS)*, 2018, pp. 1–7. https://doi.org/10.1109/ISDFS.2018.8355340.
21. K. M. Khan, J. Arshad, and M. M. Khan, "Investigating performance constraints for blockchain based secure e-voting system," *Future Gener. Comput. Syst.*, vol. 105, pp. 13–26, 2020.
22. R. Widayanti, Q. Aini, H. Haryani, N. Lutfiani, and D. Apriliasari, "Decentralized Electronic Vote Based on Blockchain P2P," *2021 9th International Conference on Cyber and IT Service Management (CITSM)*, Bengkulu, Indonesia, 2021, pp. 1–7. https://doi.org/10.1109/CITSM52892.2021.9588851.
23. K. G. Chaudhari, "E-voting system using proof of voting (PoV) consensus algorithm using block chain technology," *Int. J. Adv. Res. Electr. Electron. Instrum. Eng.*, vol. 7, no. 11, pp. 4051–4055, 2018.
24. A. Yousif, K. Rajesh, and W. Wang, "A Survey of Blockchain-Based on E-voting Systems," *ICBTA 2019: Proceedings of the 2019 2nd International Conference on Blockchain Technology and Applications*, Dec. 2019, pp. 99–104.
25. F. Þ. Hjálmarsson, G. K. Hreiðarsson, M. Hamdaqa, and G. Hjálmtýsson, "Blockchain-Based E-Voting System," *2018 IEEE 11th International Conference on Cloud Computing (CLOUD)*, San Francisco, CA, USA, 2018, pp. 983–986.
26. L. V. Thuy, K. Cao-Minh, C. Dang-Le-Bao, and T. A. Nguyen, "Votereum: An Ethereum-Based E-Voting System," *2019 IEEE-RIVF International Conference on Computing and Communication Technologies (RIVF)*, 2019, pp. 1–6. https://doi.org/10.1109/RIVF.2019.8713661.
27. S. K. Vivek, R. S. Yashank, Y. Prashanth, N. Yashas, and M. Namratha, "E-Voting Systems using Blockchain: An Exploratory Literature Survey," *2020 Second International Conference on Inventive Research in Computing Applications (CIRCA)*, 2020, pp. 890–895. https://doi.org/10.1109/ICIRCA48905.2020.9183185.

28. U. Jafar, M. J. Ab Aziz, and Z. Shukur. "Blockchain for electronic voting system—Review and open research challenges," *Sensors*, vol. 21, no. 17: 5874, 2021. https://doi.org/10.3390/s21175874.
29. H. Yi, "Securing e-voting based on blockchain in the P2P network," *EURASIP J. Wirel. Commun. Netw.*, vol. 137, pp. 1–8, 2019.
30. K. M. Khan, J. Arshad, and M. M. Khan, "Empirical analysis of transaction malleability within blockchain-based e-voting," *Comput. Secur.*, vol. 100, Jan. 2021. https://doi.org/10.1016/j.cose.2020.102081.
31. U. Jafar, M. J. Ab Aziz, Z. Shukur, and H. A. Hussain, "A systematic literature review and meta-analysis on scalable blockchain-based electronic voting systems," *Sensors*, vol. 22, no. 19, p. 7585. https://doi.org/10.3390/s22197585.
32. A. Alshehri, M. Baza, G. Srivastava, W. Rajeh, M. Alrowaily, and M. Almusali, "Privacy-preserving e-voting system supporting score voting using blockchain," *Appl. Sci.*, vol. 13, no. 2, p. 1096. https://doi.org/10.3390/app13021096.

10 Design of Intelligent Healthcare Information System Using Data Analytics

P. Nagaraj, V. Muneeswaran, and Pandiaraj Annamalai

10.1 INTRODUCTION

Data analytics is the group and analysis of data, which is produced in the healthcare industry and is useful to gain better insights and support decision-making. Real-time virtual signs can be transmitted to Electronic health records (EHR). Big data analytics are useful to obtain better knowledge about healthcare organizations with a variety of initiatives, including the surveillance of disease and the efforts of its prevention, the evolution of diagnostic and clinical methods, and the evolution of personalized, effective healthcare marketing campaigns. Implementation of services, to recognize the spreading of diseases earlier, causes new insights into disease surveillance, monitoring the quality of medical and healthcare organizations as well as coming up with better treatment methods. Healthcare data management is the process of stockpiling, protecting, and analyzing the data pulled from diverse sources [1]. Controlling the wealth of available healthcare data allows healthcare systems to create integrated views of patients, customize treatments, improve transmission, and magnify health outcomes. Data analytics which is based on the healthcare system can derive major insights into the healthcare system's waste of resources; maintain individual practitioner performance, track the health of the population, and identify people who are at risk of chronic disease.

Data analytics in the healthcare system can be treated as applying mathematical tools to a large amount of data to keep it in perfect form to help patients make informed decisions about their health. Figure 10.1 shows the four main types of data analytics that play a major role in the healthcare system and hospitalization.

They are as follows:

- Descriptive analytics.
- Diagnostics analytics.
- Predictive analytics.
- Prescriptive analytics.

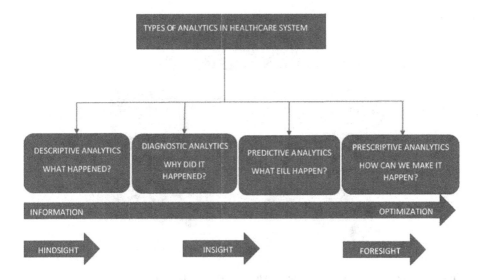

FIGURE 10.1 Different types of data roles in the healthcare system.

The above four analytics are the main sectors in data analytics that play a major role in the healthcare system and hospitalization.

10.2 WORK OF DESCRIPTIVE ANALYTICS IN HEALTHCARE SYSTEMS

Descriptive analytics is the first step that takes place in a healthcare system based on data analytics. Descriptive analytics is based on the question that the doctor is going to ask the patient, "what happened?". Descriptive analytics examines the patient's "what happened?" to him/her as well as providers who are engaging in the healthcare system. For example, suppose a person is suffering from a health issue and comes to the hospital. Figure 10.2 shows the register will get to ask the patient "what happened?". There, the person must tell the receptionist "what happened?" to him/her. After receiving the OP sheet from the reception counter, the person will get to the doctor, there the doctor will ask the patient "what happened?" to you, what health issue does he/she suffering from?

Descriptive analytics starts with the question "what happened?". This is the first step in the healthcare system starts with.

10.3 WORK OF DIAGNOSTICS ANALYTICS IN HEALTHCARE SYSTEM

Diagnostic analytics is based on advanced analytics that inspects data or computing to answer the question "why did it happen?". It is identified by the skills such as drill-down, data discovery, and data mining [2].

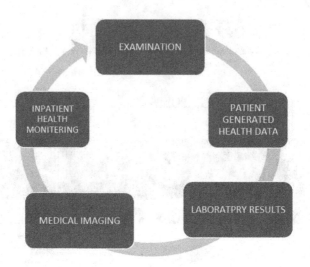

FIGURE 10.2 Examining the patient "what happened?" to him/her.

Diagnostic analytics works based on the symptoms to indicate the source of "what has happened?". The main thing is that physicians and other caretakers continue to be answerable for the final diagnosis; they can implement the data analytics strategies to save time and to keep away from possible errors of judgment. For example, we will continue the same example that we used in descriptive analytics. The diagnostic analytics begin after the doctor has questioned the patient. After explaining the patient about his/her health issue to the doctor, the doctor starts the diagnosis of the patient, which is called "diagnostic analytics." The second step starts after the explanation of the health issue by the patient to the doctor. Then the doctor starts the diagnosis of the patient "why did it happen?". By taking descriptive analytics, the doctor will give the prescription according to the diagnostic analysis. Figure 10.3 shows according to the diagnosis sample, the doctor will give the perfect prescription according to the health issue of the patient. Overall, diagnostic analytics answers "why did it happen?". Diagnostic analytics can be described mainly by various techniques including drill-down, data discovery, data mining, and correlations [3].

10.3.1 WORK OF DRILL-DOWN

Drill-down is the technique that allows the person to go in-depth into more accurate layers of the data or facts being analyzed. Typically, the format and appearance of every level in the report are consistent, but the level of detail in the data varies. By using the technique, we can create a visualization using an existing hierarchy and can convert the table to a matrix or chat form to analyze the strategy.

The use of drill-down in hospitalization helps to explore the several dimensions of the data by computing from one level down to a more detailed level. This allows the patient to get better knowledge to view aggregated, summary data and then hierarchically explore extending far down from the top surface levels of the data for more biological analysis.

Design of Intelligent Healthcare Information System Using Data Analytics 187

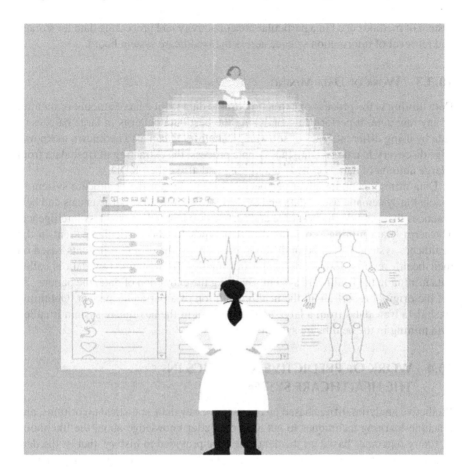

FIGURE 10.3 Examining the patient "why it happened?".

10.3.2 Work of Data Discovery

Data discovery is defined as the collection and analysis of various data from various sources to get better knowledge about awareness from hidden patterns and trends of the data. It is the first stage in the full use of harnessing an organization's data to give a brief note on critical business decisions.

Data discovery is being enabled by the business intelligence tools, such as data discovery software. The 21st-century generation of business intelligence tools is sterilely authorizing healthcare system facilities around the galaxy of the world aggressively to build up ahead with ultra-enhanced planning and rendering at macro and micro levels, alike. Healthcare technology systems offer various benefits ranging from organizing and from recording medical resources to enhancing the standard of patient treatment. By proactively managing situations, healthcare providers can comply with governmental regulation requirements. This helps healthcare providers to avoid reactive responses to situations and instead take proactive measures to prevent them from occurring. Data discovery tools are cuddles for the capacity to offer a new and existing system of methods used in a particular area of activity and processing data for storage and retrieval of information science across the healthcare system board.

10.3.3 Work of Data Mining

Data mining is the process of extracting usable data from a large amount of data set of any raw data. It performs as analyzing the best data patterns in large batches of data by using different types of software. Data mining can also be known as knowledge discovery in data (KDD). In a sample process, the extracting of fresh data from a large amount of data is called data mining and also called KDD.

Data mining contains great perspective techniques for the healthcare system to enable the systematic use of data and analytics to get to know inefficiencies and best practices that provide the best care and reduce costs, based on business intelligence. The term "data mining" contains different types for the particular patient in the healthcare system. It will be in different patterns for the different patients based on their health reports. Extracting health information from simple raw data is called data mining in the healthcare system. By using the extraction of data, the patient can get the original report about the health condition of a particular patient. Obtaining fresh data (raw data) from a large amount of data in the healthcare system is called data mining in the healthcare system.

10.4 WORK OF PREDICTIVE ANALYTICS IN THE HEALTHCARE SYSTEM

Predictive analytics differs based on the usage of raw data, statistical algorithms, and machine-learning techniques to get to know better knowledge about the likelihood of future outcomes based on the data that were provided in history, that is, the data that were stored for future use (historical data). This is based on what happened to give a better prescription of what will get to happen in the future.

Predictive analytics is one of the major types of data analytics [4]. Predictive analytics is useful to the predictive data by the electronic algorithms that can create a forecast of future events in real time. The technology can pledge the healthcare system to have the ability to make use of "big data" or data analytics-vast, real-time data sets, etc., which are collected, and which are analyzed by the EHR systems to improve the ability of patient outcomes and lower the healthcare costs care. It will create the best policy ethical and legal challenges.

Predictive models make treatment suggestions that are enabled and which are designed by the developer to improve the overall outcomes of the total health population, and these suggestions may impact the physicians, ethical commitments to act on the patient to give the best individual treatment. A system that cannot avoid expressing the connection with the ability to generate optimal outcomes that can enhance the health issues of certain patients with the specified condition based on the regulations of the medical facility due to the relatively low likelihood of their satisfaction when introducing the new technology to other interested individuals who are likely to be satisfied. Predictive analytics also have intense more to concern that across a population of a huge number of patients, those who are already underprivileged.

Design of Intelligent Healthcare Information System Using Data Analytics 189

FIGURE 10.4 Techniques of predictive analytics.

Figure 10.4 depicts how the importance of predictive analytics has significantly increased in the healthcare sector over the last decade, as the size and complexity of real-time electronic datasets have grown to such an extent that commonly used data processing tools have proven inadequate. By using the advantage of EHR, it became easy to implement predictive analytics in healthcare systems. Using predictive analytics in the healthcare system can be useful to calculate the large amount of data that are stored historically from decades of work in statistics, computer science, and clinical decision support. In this prominent era of data analytics, the use of predictive analytics plays a major role in obtaining a variety of current or historical information such as claims, and clinical, social, and genomic data to make the best predictions.

The landscape of predictive analytics is rapidly evolving, making it increasingly difficult to forecast how technological advancements will impact the reactions of physicians and patients. So, for these types of main reasons, predictive analytics models must be the same at all the subsequent times of evaluated, updated, implanted, and

FIGURE 10.5 Checking the patient data by taking the predictive analysis.

reevaluated. Figure 10.5 shows the process can be implemented by involving not only model designers but also everyone's support with the prospective legal and ethical issues embossed by the technology.

10.5 WORK OF PRESCRIPTIVE ANALYTICS IN THE HEALTHCARE SYSTEM

Prescriptive analytics relies on the utilization of machine learning algorithms to determine the appropriate action to be taken in a business context based on the predictions made by the computer program [5]. Prescriptive analytics also works along with predictive analytics, which uses raw data to get better knowledge about near-term outcomes [6]. Prescriptive analytics enables healthcare system decision-makers to get the business outcomes in the simplest form by recommending the better option of action for patients or providers. They can also think on their own to enable the comparison of multiple "what is" plans to get to access the impact of choosing one step over another. The last stage of analytics which is produced in the healthcare system is prescriptive analytics. Prescriptive analytics differ from predictive analytics, but it does not stop there to show the likely outcomes between the two analytics [7], but instead continues to reveal recommended actions to make healthcare providers more successful commercially or ecclesiastically to meet patient needs [8].

Prescriptive analytics can be used to analyze if there will be an increase in ophthalmology claims in the next year. It will help the people if they should keep the minimum ophthalmology reimbursement rates the same, increase, or decrease them by the next year. It helps everyone to make more informed decisions. Prescriptive analytics is also useful to prepare healthcare system organizations for future and

Design of Intelligent Healthcare Information System Using Data Analytics 191

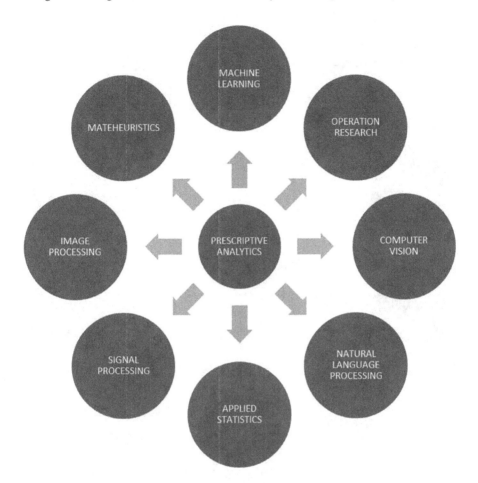

FIGURE 10.6 Prescriptive analysis techniques based on the healthcare system.

unforeseen events. Prescriptive analytics will vary from predictive analytics because it will give more benefits to the healthcare system that usually uses it. Figure 10.6 shows that predictive analytics helps the healthcare system to forecast its outcome, but prescriptive analytics digs it to take the perfect analytic actions on those findings. It helps the healthcare system to empower the efficiency of the results outcomes. Prescriptive analytics deletes all the mind guesswork in the decision-making process and optimizes it to get better improvement in the care of patients. In the end, it is all about the healthcare system delivering the best possible care to the patients.

10.6 DATA COLLECTION IN THE HEALTHCARE SYSTEM

The act of gathering data involves organizing and analyzing healthcare system data in a structured manner. It is crucial to interpret this data in order to develop, implement, and assess prevention strategies for public health. Many large countries need to enhance their information systems [8].

FIGURE 10.7 Data collection process is done in the healthcare system.

Figure 10.7 shows data collection in the healthcare system is used to create a holistic view of patients, personalized treatments, advance treatment methods, and improve socialization and communication between doctors and patients to get to know the perfect outcomes of health issues of a particular patient.

Various data collection methods were used in the healthcare system to collect data, including healthcare system surveys, administrative enrollment, and billing records. Medical records are used by a variety of entities, including hospitals, community health centers, physicians, and health plans.

10.7 DATA EXTRACTION IN THE HEALTHCARE SYSTEM

Data extraction is one of the widely used techniques in the healthcare industry. From the beginning stages of preventive support, it has been facing a critical struggle in data genetics. Figure 10.8 shows that data genetics is the best method of collecting data at a specific system at a time.

Design of Intelligent Healthcare Information System Using Data Analytics 193

FIGURE 10.8 Data extraction is done in the healthcare system.

Data extraction tries to make use of apropos data of patients that includes pharmaceutical history and comprehensive data easily accessible to approved users like the doctor [9]. It broadcasts the importance of keeping particulars secure and obtain to stop any illegal access [10]. It creates automated analytical history by including demographics, disease history, preventive checkups, or fitness checkups of all the cases. Nowadays, data extraction in healthcare systems takes place widely in educational circumstances [11].Getting it into health plans and making **significant** changes **requires** three methods. They are analytics, best practice, and adoption, along with an antecedent of evolution.

10.8 DATA GENERATION IN THE HEALTHCARE SYSTEM

Data generation in the healthcare system is heroic and rapid, involving flexibility in an amalgamation of methods for secure data storage and computing [12].

Figure 10.9 shows the clause that data analytics has been introduced to describe a huge amount of data sets that exceed the volume of typical hardware properties to manipulate, store, and analyze [13].

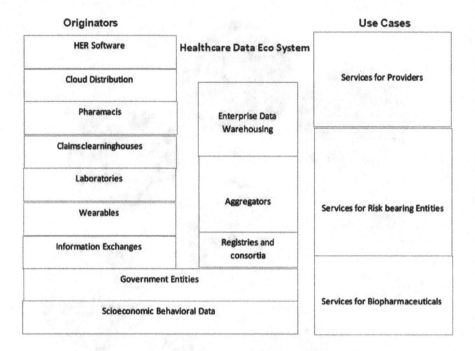

FIGURE 10.9 Data generation is done by the healthcare system.

10.9 ANALYSIS OF THE HEALTHCARE SYSTEM

Analysis works on the principles that focus on insights into hospital management, patient records, costs, diagnosis, and much more done by using the technique of analysis in the healthcare system. The analysis covers a lengthy sociable of the healthcare system, providing perceptions on both the macro and micro participation level [14]. Figure 10.10 shows healthcare analysis is based on the system of data analysis, which permits healthcare system experts to procure possibilities for amelioration in healthcare industry management, patient appointment, pay, and diagnosis [15].

10.10 VISUALIZATION AND REPORTING IN THE HEALTHCARE SYSTEM

Data visualization provides the best important techniques in the healthcare system focus; it is very useful to identify the important patterns and correlations and provides data analysis more systematic [16]. Data visualization is one of the best institutes for health metrics and gauging is just one of the main important organizations sharing the visualizations [17].

Incident reporting (IR) in the healthcare system has been prescribed as important to improve the specialization of patient safety. The intent of IR is to recognize the best safety threats and expand interventions to alleviate these threats to decrease immorality in the healthcare system [18]. Dashboards are useful to help medical

Design of Intelligent Healthcare Information System Using Data Analytics 195

FIGURE 10.10 Analysis of data done by the healthcare system.

professionals analyze the huge amount of data sets quickly to save time and even to save the lives of patients by the medical professionals [15]. The evaluation of a dashboard to visualize EHL created a 65% depletion in time spent on data analysis during the year alone, with further gains projected for the future.

A huge amount of healthcare organizations have already started providing the data in the forms of graphs and charts that are made by the customer for the presentation of raw data. Figure 10.11 shows that visualization is much more interactive and customizable in proportion to the data than the viewers wish to get better knowledge on health. The data with which you are working and the respected people you are targeting will both determine the best types of evolving medium and data visualization to use [19]. If the work is in any case, however, data visualizations build champing huge amount of data set to be more efficient and produce the best perceptions as a result. Visualization of data analysis helps professionals find the best techniques to analyze the health of the patient in the healthcare system and gives the best decisions according to the health analysis of the patient [20].

10.11 PROBLEMS, CHALLENGES, BARRIERS, AND ISSUES IN THE HEALTHCARE SYSTEM

We are living in a digitalized world; we are producing a large amount of data every minute. The data that were produced by the users at every minute makes it challenging to store, manage, utilize, and analyze it [21]. Even a lot of business management

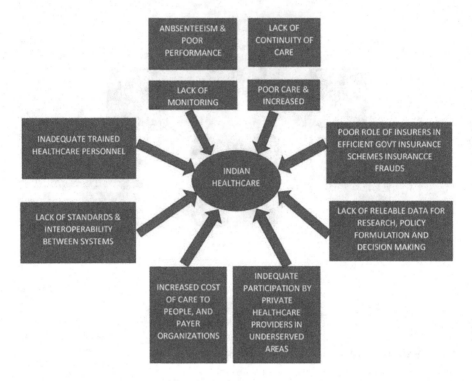

FIGURE 10.11 Problems and challenges faced in the healthcare system.

systems are struggling to get a way to make this huge amount of data usage. Every year, 40%–60% of data are not useful to the professionals; they want the useful data of the patient to make use of analytics. A greater number of organizations are utilizing data capture and big data analytics tools to handle the data by preparing useful data to a great extent [22].

Moreover, the major challenges faced by doctors and professionals are as follows:

1. Need for the operation that was done more than once or two quickly across disparate data sources.
2. Drastic scarcity of professional doctors who understand big data analysis.
3. Producing useful insights utilizing the use of big data analytics.
4. Producing commodious data into the big data platform.
5. The unpredictability of data management landscape.
6. Storing the huge amounts of useful data and the quality of that useful data.
7. Securing useful data and producing privacy to that user data.

In Figure 10.11 to overcome these analytical challenges in the healthcare system organizations, they should be more accurate in the training programs to analyze the big data analytics in the healthcare system [23].

FIGURE 10.12 Management issues faced by the healthcare system.

10.12 MANAGERIAL ISSUES IN THE HEALTHCARE SYSTEM

Figure 10.12 shows how to dissect the monetary medical services framework, which is incorporated into parts, for example, dealing with finance, patient information passage, handling guaranteeing of protection or clinical utilizing coding administrations, clinical charging and refusal by the executives, and course of action for reimbursement.

The major issues are based on the following:

1. Costs and transparency.
2. Consumer experience.
3. Delivery system transformation.
4. Data and analytics.
5. Interoperability of consumer data access.
6. Holistic individual health.
7. Related: the future of healthcare leadership.
8. Next-generation payment models.

10.13 DATA QUALITY IN THE HEALTHCARE SYSTEM

Data quality contains techniques in the healthcare system based on general characteristics, including precision, stability, and relevancy [24]. By using data quality, healthcare information technology cheer-ups the quality of data by lowering redundancy, decreasing medical errors, and enhancing health outcomes.

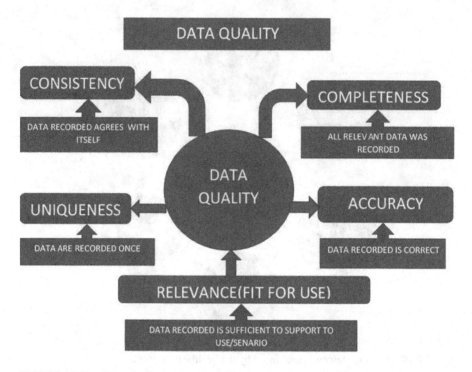

FIGURE 10.13 Data quality is done in the healthcare system.

Healthcare systems must produce quality data and have to build up a strong base to manage it in a long term in terms of a concept in a structured manner. By producing those, we can expect the speed up of existing process and build learnings that provide smarter policy decisions that can hold on all stakeholders.

Figure 10.13 shows that data quality can give the best quality the picture of what others can see in your data and the use of portable data storage's attainable to other patients or doctors or professionals [25]. We should always explore our raw data to learn about health issues clearly. Set up a single-grade report system based on data quality metrics and coding.

10.14 PUBLIC REPORTING DATA IN THE HEALTHCARE SYSTEM

Public reporting is widely used to get to know about the publishing of information and reports on the quality of care in the healthcare system and long-time care providers for making use of patients and users [26]. This is mostly developed on social platforms and the Internet, it is also published in magazines and in other formats to develop healthcare system facilities. Not only by aiding patients and users to get knowledge about choices but also by public reporting systems are targeted in different ways to give the best reports of certain healthcare systems to a different audience [27]. Quality is being initiated in different ways. Figure 10.14 shows different reports of health and long-term care meaning that these different ways might be given different importance in different sectors [28].

Design of Intelligent Healthcare Information System Using Data Analytics 199

FIGURE 10.14 Examples of public reporting in the healthcare system.

Queries and guidelines can be instructed to indicators to encourage their contributions and reporting, but also there is a concern that they can precipitation unfavorable providers' performance. To increase public reporting, there are some techniques that the healthcare system should follow:

1. Exhibiting pertinent information.
2. Calculating measures to match the skill levels of users.
3. Upgrading demonstration methods.
4. Instructing patients and users about quality in health and long-term care and expanding patient and user consciousness of public reporting.
5. Engaging professionals in keeping up public reporting systems.
6. Scheming decision aids and supporting their use.

Therefore, when considering how to apply public reporting, specifics consider the following:

1. Monetary and human facility necessity.
2. Whether reporting will be compulsory or discretionary for caretakers.
3. Whether data might be produced as a part of regulatory surveys or clinical inspections.
4. Techniques for addressing poor production in data supply.

10.15 DATA PRIVACY AND GOVERNANCE IN THE HEALTHCARE SYSTEM

The health insurance portability and accountability act (HIPAA) is a US-based law taken in 1966 for data security regulation. HIPAA helps the data to get a clear definition of what inaugurates health data or "protected health information" (PHI). Health insurance

FIGURE 10.15 Data governance and security did in the healthcare system.

portability and accountability act make use of 18PHI identifiers names, zip codes, medical records, social security number, etc. Securing privacy can commerce more effectual communication between physician and patient, which is crucial for the quality of the case, intensifying independence, and averting economic harm, awkwardness, and prejudice. The patient who is suffering from health issue should be confidential about the pillars of medicine [29]. Securing the personal details of a patient who is suffering from a health issue is just not a matter of virtuous respect; it is necessary in maintaining the important bond of trust between the doctor and the patient keeping up privacy and confidentiality help to take care of the patient to protect from potential harms including psychological harm such as embarrassment or distress; social based harms like loss of employment or damage to one's financial standing; and criminal and civil liability [30].

Figure 10.15 shows that data governance in healthcare is perturbed about how to patient, secure, and precisely gather each piece of data [31]. Statistics governance in the healthcare system is an important process and system to maintain the data to make decisions about patient care. Data governance in the healthcare system helps to secure unofficial access to patients' private health information [32]. Personal health information (PHI) was clearly defined by the data governance to set strategies that safeguard health data and keep it safe.

10.16 CONCLUSION

Based on the results produced from the data analytics, specified steps or measures can be taken to reduce the general costs of the healthcare facilities provided in the healthcare system. Decreased costs may be thrown back in the form of decreased

Design of Intelligent Healthcare Information System Using Data Analytics 201

expense freight on the customers of those healthcare facilities. By analyzing the data, which were produced in the healthcare system, the patient will get more useful prescriptions according to their health issue. By employing a variety of analytics that produce perfect data to examine a specific patient's health, the healthcare system can improve its understanding of a patient's health by applying the principle of four analytics—descriptive analytics, diagnostic analytics, predictive analytics, and prescriptive analytics. The overall chapter explains how data play a major role in healthcare analytics based on the types of operations done in the healthcare system.

REFERENCES

1. C. El Morr, and H. Ali-Hassan, "Healthcare, data analytics, and business intelligence," In: *Analytics in Healthcare*, Springer, Cham, pp. 1–13, 2019.
2. V. Muneeswaran, P. Nagaraj, U. Dhannushree, S. Ishwarya Lakshmi, R. Aishwarya, and B. Sunethra, "A framework for data analytics-based healthcare systems," In: Raj, J.S., Iliyasu, A.M., Bestak, R., Baig, Z.A. (eds) *Innovative Data Communication Technologies and Application*. Lecture Notes on Data Engineering and Communications Technologies, vol. 59, Springer, Singapore, 2021. https://doi.org/10.1007/978-981-15-9651-3_7
3. M. Aiello, et al., "The challenges of diagnostic imaging in the era of big data," *J. Clin. Med.*, vol. 8, no. 3, p. 316, 2019.
4. B. Van Calster, et al., "Predictive analytics in health care: How can we know it works?," *J. Am. Med. Inform. Assoc.*, vol. 26, no. 12, pp. 1651–1654, 2019.
5. A. Belle, R. Thiagarajan, S. M. R. Soroushmehr, F. Navidi, D. A. Beard, and K. Najarian, "Big data analytics in healthcare," *BioMed Res. Int.*, vol. 2015, Article ID 370194, 2015.
6. J. Lopes, T. Guimarães, and M. F. Santos, "Predictive and prescriptive analytics in healthcare: A survey," *Procedia Comput. Sci.*, vol. 170, pp. 1029–1034, 2020.
7. M. Viceconti, P. Hunter, and R. Hose, "Big data, big knowledge: Big data for personalized healthcare," *IEEE J. Biomed. Health Inform.*, vol. 19, no. 4, pp. 1209–1215, 2015.
8. G. Tripathi, M. A. Ahad, and S. Paiva, "S2HS-A blockchain based approach for smart healthcare system," *Healthcare*, vol. 8, no.5. Elsevier, 2020.
9. P. R. Jeyaraj, and E. R. S. Nadar, "Atrial fibrillation classification using deep learning algorithm in internet of Things–based smart healthcare system," *Health Inform. J.*, vol. 26, no. 3, pp. 1827–1840, 2020.
10. X. Ma, Z. Wang, S. Zhou, H. Wen, and Y. Zhang, "Intelligent healthcare systems assisted by data analytics and Mobile computing," *Wirel. Commun. Mob. Comput.*, vol. 2018, Article ID 3928080, p. 16, 2018. https://doi.org/10.1155/2018/3928080
11. S. Altowaijri, R. Mehmood, and J. Williams, "A quantitative model of grid systems performance in healthcare organisations," in *Proceedings of the 2010 International Conference on Intelligent Systems, Modelling and Simulation*, pp. 431–436, January 2010.
12. M. Chen, J. Yang, Y. Hao, S. Mao, and K. Hwang, "A 5G cognitive system for healthcare," *Big Data Cogn. Comput.*, vol. 1, no. 1, p. 2, 2017.
13. V. Subramaniyaswamy, et al., "An ontology-driven personalized food recommendation in IoT-based healthcare system," *J. Supercomput.*, vol. 75, no. 6, pp. 3184–3216, 2019.
14. M. Shahbaz, et al., "Investigating the adoption of big data analytics in healthcare: The moderating role of resistance to change," *J. Big Data*, vol. 6, no. 1, p. 6, 2019.
15. C. L. Ventola, "Mobile Devices and apps for health care professionals: Uses and benefits," *Pharm. Therap.*, vol. 39, no. 5, pp. 356–64, 2014.
16. S. Shafqat, et al., "Big data analytics enhanced healthcare systems: A review," *J. Supercomput.*, vol. 76, no. 3, pp. 1754–1799, 2020.

17. M. M. Hassan, K. Lin, X. Yue, and J. Wan, "A multimedia healthcare data sharing approach through cloud-based body area network," *Future Gener. Comput. Syst.*, vol. 66, pp. 48–58, 2017.
18. A. M. Rahmani, T. N. Gia, and B. Negash et al., "Exploiting smart e-Health gateways at the edge of Healthcare Internet-of-Things: A fog computing approach," *Future Gener. Comput. Syst.*, vol. 78, pp. 641–658, 2018.
19. M. Chen, Y. Ma, Y. Li, D. Wu, Y. Zhang, and C.-H. Youn, "Wearable 2.0: Enabling human-cloud integration in next generation healthcare systems," *IEEE Commun. Mag.*, vol. 55, no. 1, pp. 54–61, 2017.
20. G.-H. Kim, S. Trimi, and J.-H. Chung, "Big-data applications in the government sector," *Commun. ACM*, vol. 57, no. 3, pp. 78–85, 2014.
21. C.-H. Lin, L.-C. Huang, S.-C. T. Chou, C.-H. Liu, H.-F. Cheng, and I.-J. Chiang, "Temporal event tracing on big healthcare data analytics," in *Proceedings of the 3rd IEEE International Congress on Big Data, BigData Congress 2014*, pp. 281–287, USA, July 2014.
22. A. Darwish, et al., "The impact of the hybrid platform of internet of things and cloud computing on healthcare systems: Opportunities, challenges, and open problems," *J. Ambient Intell. Humaniz. Comput.*, vol. 10, no. 10, pp. 4151–4166, 2019.
23. K. Lee, T. T. H. Wan, and H. Kwon, "The relationship between healthcare information system and cost in hospital," *Pers. Ubiquitous Comput.*, vol. 17, no. 7, pp. 1395–1400, 2013.
24. H. Zhang, S. Mehotra, D. Liebovitz, C. A. Gunter, and B. Malin, "Mining deviations from patient care pathways via electronic medical record system audits," *ACM Trans. Manag. Inform. Syst. (TMIS)*, vol. 4, no. 4, pp 1–20, 2013.
25. R. Mehmood, M. A. Faisal, and S. Altowaijri, "Future networked healthcare systems: A review and case study," In: *Handbook of Research on Redesigning the Future of Internet Architectures*, pp. 531–555, 2015. https://doi.org/10.4018/978-1-4666-8371-6.ch022
26. P.-Y. Wu, C.-W. Cheng, C. D. Kaddi, J. Venugopalan, R. Hoffman, and M. D. Wang, "-Omic and electronic health record big data analytics for precision medicine," *IEEE Trans. Biomed. Eng.*, vol. 64, no. 2, pp. 263–273, 2017.
27. R. Nambiar, R. Bhardwaj, A. Sethi, and R. Vargheese, "A look at challenges and opportunities of Big Data analytics in healthcare," in *Proceedings of the 2013 IEEE International Conference on Big Data, Big Data 2013*, pp. 17–22, USA, October 2013.
28. V. Muneeswaran, M. P. Nagaraj, M. P. Rajasekaran, N. S. Chaithanya, S. Babajan, and S. U. Reddy, "Indigenous health tracking analyzer using IoT," in *2021 6th International Conference on Communication and Electronics Systems (ICCES)*, Coimbatre, India, pp. 530–533, 2021. https://doi.org/10.1109/ICCES51350.2021.9489052
29. P. B. Prince, and S.P. J. Lovesum, "Privacy enforced access control model for secured data handling in cloud-based pervasive health care system," *SN Comput. Sci.*, vol. 1, no. 5, pp. 1–8, 2020.
30. A. Pandiaraj, S. L. Prakash, and P. R. Kanna, "Effective heart disease prediction using hybridmachine learning," in *2021 Third International Conference on Intelligent Communication Technologies and Virtual Mobile Networks (ICICV)*, pp. 731–738, 2021. https://doi.org/10.1109/ICICV50876.2021.9388635
31. V. Chandola, S. R. Sukumar, and J. Schryver, "Knowledge discovery from massive healthcare claims data," in *Proceedings of the 19th ACM SIGKDD International Conference on Knowledge Discovery and Data Mining, KDD 2013*, pp. 1312–1320, USA, August 2013.
32. B. D. Deebak, et al., "An authentic-based privacy preservation protocol for smart e-healthcare systems in IoT," *IEEE Access*, vol. 7, pp. 135632–135649, 2019.

11 Automatic Room Light Controller Using Arduino and PIR Sensor

Huma Khan, Harsh Dubey, and Yasir Usmani

11.1 INTRODUCTION

People are increasingly looking for automation in their daily lives these days. Surprisingly, people are now motivated to conserve the energy used in daily life. They are less likely to turn off the lights as they exit a room. Therefore, if the light is ON without a person, a tremendous amount of energy is wasted. Most people are generally not interested in turning off consumer electronics like fans, lights, and other office equipment when they are gone, especially in open spaces and private offices. As consumer electronics and household appliances are used more frequently and grow in size, power consumption in the home will inevitably increase. In addition, irrational power consumption occurs in public and private spaces without anyone present. Robotic replacement of the home's lighting system will result in significant energy savings, which will allow the home's owner to save money. People are currently looking for robotization to complete all of their daily chores. People are making efforts to minimize human endeavors. The automated replacement of the household lighting system significantly minimizes human labor. If the person uses programmed exchanges, they won't have to give any thought to putting the lights out before they leave the room. This mechanism also aids in decreasing the amount of electricity that is wasted energy when the fans, lights, as well as other electric appliance devices, are still ON even though there is no sign of a person being there. Infrared (IR) sensors are used as the component for person recognition [1].

As consumer electronics and household appliances are used more frequently and grow in size, electricity consumption in the domestic sector will inevitably rise. Additionally, unusable power usage occurs in both public and private spaces when no one is there. Through the utilization of robots to replace the lighting system in a house or office, the owner can significantly reduce the amount of power used while also saving money. People are already looking forward to more automation in their everyday lives. People are making efforts to reduce human endeavors. By utilizing the suggested framework, energy waste can be reduced because, with the aid of a passive infrared (PIR) sensor, electrical equipment will be turned ON or OFF automatically depending on the presence of a person. This eliminates the need to turn OFF equipment before leaving or turn it ON when you arrive at your lodge. The anticipated structure has essentially been improved in this way. Figure 11.1 shows the board of Arduino UNO. Three essential elements make up the suggested

DOI: 10.1201/9781003353034-11

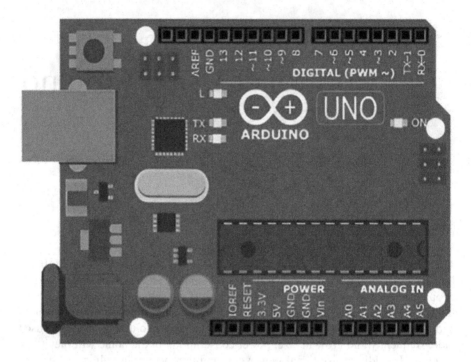

FIGURE 11.1 Arduino UNO.

structure: the Arduino, PIR sensor, and Relay Module. This system makes extensive use of PIR sensors [2, 3].

The future of the Internet of Things (IoT), a rapidly expanding technique is infinite because data streams have doubled over time. Upcoming businesses will be transferred from Cloud computing as well as Big Data Analytics, which were largely driven by the Covid-19 outbreak, have replaced more traditional data processing methods. This shift is taking place as companies everywhere adopt cloud-based working practices. Fish require their owners' sufficient consideration and care because they are considered to be quite sensitive to their surroundings. Because people frequently overlook changing the water, feeding the fish, and checking the pH levels, the IoT can simply address these issues and other simple but challenging tasks like tracking hundreds of fish. For modest- to large-scale aquariums, such as the Dubai Aquarium, it is vital to have an IoT system (where hundreds of fishes are constantly monitored). It is possible to employ the system to establish the perfect circumstances needed for high yield. Due to the global fish shortage, the aquaculture industry will be extremely important to the economy in the future. Processes must be streamlined in order to boost efficiency and enhance fish health. Ocean wildlife can be saved via IoT-based aquariums by developing dependable systems with real-time data processing capabilities. In distant areas, enormous tanks can be constructed for endangered animals, boosting biodiversity and creating a healthy ecology. Since it might be challenging to monitor larger fishes in huge tanks, technologies such as Radio Frequency Identification (RFID) and transmitters can be used to keep an eye on them [4].

Independent of the hardware and operating system employed, all kinds of intelligent equipment can communicate with one another, and the data collected through the sensors could be maintained in a cloud platform and examined for predictive analysis and alternative preservation. The water levels inside the aquarium can be recorded by the IoT-based aquatic surveillance system, and when they fall, an email can be sent to the user [5]. It is indeed capable of controlling the automatic feeder, recording thermodynamics and moisture content readings, and turning ON and OFF the aquarium's lighting with the help of Amazon Web Services (AWS) technology. The user can operate the fish feeder by either a web/app interface or a voice application. Heating, moisture, level of water, condition of LEDs, and other information about the feeder are among the parameters employed in this project. The ESP-32 performs sensor collection and serves as a local server and controller as well as a data processing device. As long as there is an internet connection, because of processing of data is done by AWS Lambda, which is recorded in DynamoDB as well as served via AWS Application Programming Interface (API), the consumer may monitor the conditions of the aquarium either local or remote location just about anywhere in the world. For effective decision-making, using Quick Sight and AWS IoT Analytics, data is further visualized. This model's features all operate smoothly and with great accuracy. This project has numerous industrial applications, including fish farming, fish management, aquaponics, and zookeeping. The knowledge is saved for further research and may be utilized to complete relevant tasks. They are very helpful for maintaining a balanced ecology and for farmers who raise particular species of fish [6].

11.2 LITERATURE SURVEY

The project's ultimate goal is to use ESP32 to power electrical appliances. With the use of smartphones, people may exercise whether or not there is an internet connection by making our workplace more opulent and intelligent. In order to operate electrical devices utilizing our mobile devices to control things like lighting and fans and IoT mobile applications, a new methodology has been devised. For those who need to make their homes smarter, it offers an easier integration and less expensive alternative. The switch that has to be created is a modular plug-and-play switch. Relay, MCB, LED indication, and two-way switch are all components of the switch board. Through the ESP32, which has built-in Wi-Fi and Bluetooth, mobile devices are connected to the switch, allowing us to operate the appliances with or without an internet connection. The idea of an RFID-based automatic room light controller is also presented in this study [7]. The IoT is a term that refers to a type of technology that links billions of actual physical objects to the internet. One of the cutting-edge technologies that has become prevalent in many application domains is automation. A summary of the planning and implementation of switch automation systems as one of its components has been included in this study. A straight forward GUI program, which might be used in the smart mode of operation, can be used to quickly and easily control the commands that are given to turn home appliances ON and OFF, such as the lighting and fans, which are referred to as control inputs. A large portion of our everyday activities and the gadgets we use depend on Wi-Fi, which is undoubtedly one of the most beneficial technological developments in the past 10 years. Wi-Fi is compatible with smartphones, laptops, and computers. To perform particular activities, these devices have the capability of being linked to a Wi-Fi network [8].

In order to address the shortcomings of traditional lighting control systems' single function, poor level of automation, and low management efficiency, this study develops and deploys a kind of smart visitor room control system for lighting that coordinates manual and automatic housekeeping light control. The system utilizes a multilayer distributed topology that incorporates CAN bus, network, sensor, and embedded technology. Its key component is an integrated ARM CPU, which enables the implementation of intelligent lighting control in guestrooms through the CAN bus communication protocol in conjunction with a number of modules. This chapter provides a thorough examination and justification of the hardware, software, and general structure designs. To strengthen management standards and service quality, lower-energy use, and increase the hotel's competitiveness, the Smart Hotel Room Lighting Control System competes with high-star hotels. It will be a crucial component of the Advanced Hotel System and has a particular use value [9].

The behavior of dimming in a smart home (SH) lighting system that was set to a constant level of illumination is described as being verified in the chapter. The suggested measurement technique was implemented as an improvement for the analysis of daylight-induced interior illumination curves. It was measured in a reference room in SH with a technology that assessed the illumination at the locations of visual tasks and simultaneously scanned the illumination brought on by daylight on an open, horizontal surface outside. The employed measurement instrument was equipped with a dimmable artificial lighting system that used a luminance controller mounted on the room's ceiling in conjunction with the KNX control system to maintain a level that remains constant illumination despite the impact utilizing both artificial and natural sources of light. The chapter demonstrates that there may be a discrepancy between the illuminance set value, controlled by a KNX system, and actual illuminance at the location of a visual task. This discrepancy could cause the apparatus for producing light to fall short of normative standards, which is totally unacceptable [10].

The planning and development of an autonomous constitution for the control room conditions are presented in this chapter. The lights and the fan will switch ON automatically whenever there are occupants in the room according to the appropriate temperature for the workstation. If there are no occupants in the room, the lights and fan will not turn ON. PIR sensor and LM35 are used in this study as input, and an FPGA-based component was created. As only one slice of the FPGA is utilized in this design, the IP core of the component is ready for fabrication. One of the key factors in promoting human comfort is automation; a technology that does tasks automatically makes them comfortable and eliminates the need for physical labor. In addition to being simple, it has favorable effects on users including a significant reduction in energy use as every task is performed by machines using accurate programming [11].

All activities and items used in daily life cannot be completely isolated from technology due to the acceleration of technological growth in the fields of science and technology. An outdated system is replaced with a brand-new one as a result of this update. Microcontroller technology, which is now applicable to every part of the environment, is one trending development. The State Electricity Company will consume less electricity overall by creating an autonomous illumination design capable of conserving power in a room (PLN). In this prototype method, the Arduino UNO microcontroller serves as the sensor's control hub and integrates a PIR sensor for human movement detection. In addition to that, there is a light-dependent resistor called (LDR) sensor in addition to the PIR sensor known as a flash sensing element. Its purpose is to determine the amount of light that is required in the room, allowing controlled lighting to be used in line with that amount of light. A GSM Sim 900A is also there, and it serves as a notification system in the case that electricity is used elsewhere other than the sensor controller [12, 13].

Smart rooms employ power generation management and environmentally conscious power generation to enable autonomous control and surveillance of various room equipment due to the rise in the electricity requirements for rooms. A smart room is one that provides users with high levels of comfort while using electricity wisely. This chapter makes a proposal for automated classrooms that take into account student comfort and solar energy as a renewable energy source. For the automatic management and supervision of classroom equipment, including lighting, fans, air conditioners (ACs), and projectors, WAGO programmable logic controllers (PLCs) are employed. Additionally, in order to gather data from the sensor as well as produce an outcome in response to input from a WAGO PLC, an Arduino UNO has been utilized as data collection equipment. Through a PLC Ethernet gateway, monitoring of multiple input sensors and electrical devices is also possible in real time wirelessly on a smartphone via Wi-Fi [14, 15]. Appliances in classrooms will only switch ON to create a usable atmosphere when PIR sensors detect occupancy. The sensors (temperature, lux, and PIR), Arduino UNO, solar charge controller, WAGO PLC, photovoltaic panels, and Wi-Fi router make up the room automation system [16–18].

As workers spend more time there and become more comfortable, the office environment is modified to allow intelligent control and monitoring of numerous parameters. In this work, an IoT system on the basis of a Thingspeak cloud and an arm controller is described. It provides a method for the remote monitoring of flexible workspace types of machinery, such as lighting, ACs, and fans, while also considering the comfort of users. It is also possible to monitor and analyze a variety of office space metrics through the internet, including temperature, amount of light, connected load, amount of energy used, voltage, and current use. Electrical appliances won't start unless the PIR sensors detect a presence. Using lux sensors and temperature sensors, respectively, a workable room environment is maintained and monitored, including light intensity and room temperature. The IoT-based room automation system is made up of the sensors (lux and temperature), multifunction energy meter, arm controller board, ESP8266 Wi-Fi module, and Wi-Fi router [19, 20].

In this chapter, the development and implementation of an intelligent control scheme for a room's lighting, air conditioning, and ceiling fans are discussed. This system takes users' needs into account correctly and creates a safe space for them. An Android application has been developed with an intuitive graphical user interface and straightforward interaction with a smart control system. The program can respond fast since it is linked to the control device through a cloud network and Wi-Fi [21, 22]. The requirements of the situations and lighting settings were taken into consideration when designing the RGB LED panel lighting technology, which controls luminance, color, and temperature. Despite the widespread availability of RGB LED panel lights, there are no standardized solutions for activity-based color modes. A system for controlling IR remote modules was created to allow users to issue commands through a smartphone application to control inverter-style ACs. To operate the air conditioning system, this technique averages the preferences of the users and numerous people into account. Many users at the same location choose the AC's operating mode using the information provided through the application [23, 24]. The fan does not continuously adjust the temperatures, unlike inverter type ACs. A manual and automatic ceiling fan control system was created as a solution to this issue. The user's hybrid system which is related to skin moisture is used in automatic mode to control the room's temperature. Just like the air conditioning system, the manual mode operates in accordance with recommendations made by the crowd. A control system uses a traditional remote controller for the AC as well as the ceiling fan which relies on crowd sentiment [25–27].

The chapter's objective is to show how to eliminate the needless electricity waste. Typically, when leaving a room, humans may forget to turn off the lights and fans. Even though we may not realize it, we are squandering a lot of electricity. A controller that can detect a person's presence and take the appropriate action will consume less electricity and can be installed in homes, schools, workplaces, etc. Monitoring the presence of individuals in the room is done using IR sensors, PIR sensors, and an Arduino board. Therefore, the lights and fans should ideally turn on when the room is vacant and when someone enters in, and they should switch off automatically when no one is present. Controlling electricity waste and improving people's quality of life are the main goals of this endeavor [27–30].

Humans typically receive the majority of their visual perception of the world. Consequently, light is a crucial component that enables individuals to recognize the dimensions, tone, and viewpoint of their surroundings. Furthermore, assuming that buildings are responsible for around 40% of total energy use worldwide, housing energy efficiency is a problem that is actively being researched. Additionally, it is estimated that lighting utilizes 20% of overall energy consumption in homes and 40% to 70% in businesses such as stores and offices. This study provides a fuzzy logic controller that effectively uses both natural and artificial light to preserve a meeting room's visual comfort. In order to evaluate how well the suggested control system works, the pertinent real findings generated are presented and discussed [31, 32].

A microcontroller board called the Arduino UNO has 14 digital input and output pins, 6 of which can be utilized as PWM outputs and the remaining 6 as straightforward data inputs. A button for resetting the device, a USB connector, a power jack, an ICSP header, and a crystal oscillator operating at 16 MHz are also included. Everything required to operate the microcontroller is included with a battery or a DC connector or to interface it with a PC using a USB link. The Arduino UNO and MEGA are the two most often used pre-owned Arduino boards, while there are a few other types as well. The only difference between these two is the number of information and results from pins, where the Arduino MEGA has more than the nine Arduino UNO. An Arduino, like any microcontroller, is a small tiny printed circuit board that contains a chip capable of being customized to execute a variety of jobs. Data is transferred from software running on a personal computer to an Arduino microcontroller. The Arduino microcontroller then transfers the data to the appropriate circuit or machine, which may have multiple circuits, in order to carry out the precise command [19, 33]. This is how Arduino is commonly used in microcontroller programming today in addition to other things. According to Yusuf Abdullahi Badamas, Arduino can deliver data to output devices like LED, speakers, LCD screens, and DC engines and assist with data processing from input devices like sensors, radio wires, and trimmer potentiometers. Tiffany Tang asserts that work has also been done to integrate Arduino with Kinect to control movement, and Shamsul Aizal Zulkifli, Mohd Najib Russin, and Tiffany Tang assert that Arduino can also be utilized as a microcontroller for a three-stage inverter. This demonstrates the wide range of uses for an Arduino, and in this task, Arduino will be used to manage data obtained from a pH sensor before being transmitted to a cell phone [34, 35]. The water siphon and engine will also be started using Arduino. Arduino UNO board 10 is shown in Figure 11.2.

Intelligent systems for home automation are in extremely high demand right now. These systems make it simple for users to use home appliances from anywhere and quickly. In this chapter, we introduce an SH security system based on the IoT. A smart locking mechanism is used to give the house security. Every time a user or visitor inputs a password into a digital lock, their face is photographed and sent as a message to the home's owner. If a user makes three consecutive unsuccessful attempts,

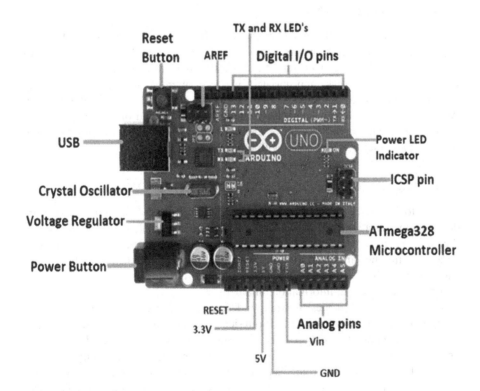

FIGURE 11.2 Arduino UNO board pin index.

an intrusion has been discovered. Based on the information obtained, the owner will decide whether to permit the user or guest admission to the house [36–38]. SHs are equipped with cutting-edge technology like sensors, wired and wireless networks, actuators, and intelligent systems. Since the early 1990s, the idea of the smart house has advanced quickly. Yet even though many people believe SHs to be a part of "future living," their popularity is still small and they have not become widely popular after over two decades. By providing architects, designers, and planners with a thorough overview of the SH, its technologies, goals, problems, and obstacles as well as suggestions on how to get around them and potential implications for how we will live in the future, this chapter aims to close the knowledge gap between the fields of computing, architecture, and health care [39].

The building blocks of smart cities are smart houses. Since our homes and apartments serve as the essential building blocks of our infrastructure, all infrastructure-related problems must be addressed by codes that direct people toward smarter housing. Additionally, these standards ought to guide us toward efficient water and waste management, green construction, and a safe and healthy living environment. This study demonstrates how various Smart city infrastructure projects are closely tied to the fields of architecture and engineering and the requirement for a robust Smart city-specific building code at the national and municipal levels [40].

11.3 IoT COMPONENTS

Devices, gateways, and the cloud are the three fundamental components of IoT, as depicted in Figure 11.2. Actuators, user interface components, and sensors are examples of devices. These gadgets have the necessary hardware and software for using the internet. In order to act appropriately, gadgets communicate with one another or a control unit. To access cloud services, gateways connect devices to the internet. Additionally, it makes device security features available and offers complete data security. To complete the given task, each device's data is transported to the cloud, which is processed and combined with other devices' data [14].

11.4 ARCHITECTURE OF ROOM AUTOMATION WITH IoT

Thanks to IoT-based room automation architecture, networking, communication, device administration, and data analysis are all customizable. The architecture of the smart room automation system has been represented in Figure 11.3. Physical, data link, middleware, and application layers are all components of the overall smart room architecture. The physical layer includes a physical node for producing, accepting, or retrieving information on a network, as well as appliance nodes, such as electric home appliances, sensors, and actuators. There will be an interaction between the application layer and the physical layer provided by the data link layer. In addition to computational analysis, the data will be stored, recovered, processed, and registered. Application programs and the data link layer can communicate via a variety of communication protocols provided by middleware. Based on a home communication network, the application software layer is in charge of providing the user with control features and monitoring modules [13, 15]. In light of IoT technology, an

Automatic Room Light Controller Using Arduino and PIR Sensor 211

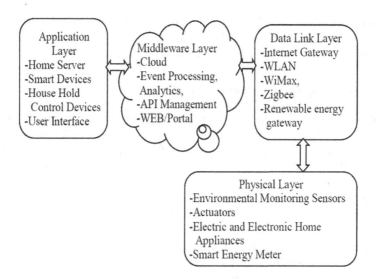

FIGURE 11.3 Architecture of room automation with IoT.

SH system architecture that incorporates house types of equipment in a smart system has been presented in Ref. [16].

11.5 PROPOSED CIRCUIT DESCRIPTIONS

The power of lights will rely on how many people are in each room for the recommended arrangement of the home lighting controller's energy-saving approach [3]. The suggested configuration is broken down into three different parts: a relay system for pre-programmed exchanges, a human indication with PIR sensors, and a microprocessor circuit for operating the complete system. For actual current flow reading, Arduino initially receives power source current, which is followed by the power indicators of numerous connected components, and then waits for commands from a few input components, specifically the proximity sensor and PIR sensor. There will either be a current flowing or the light sensor will be in the HIGH state and additional instructions redirected toward the outcome if the brightness of the room is reduced; however, the instruction is not sent until the current has ceased flowing. When an input with a HIGH value or current transmits instructions, in order to turn the lights ON, Arduino navigates to the next output, which converts the LOW value read from a relay to the HIGH value. Just after lights have been turned on, Arduino executes the following command, which is to flow the current so that it can be read by the GSM Sim 900A module. This allows Arduino to send an SMS notice to the user indicating that the light has been turned ON. Once the lights are turned OFF, an SMS will be sent letting you know that they are OFF. PIR and light sensors act as input devices, producing output values that the Arduino UNO microcontroller processes. All installed programs from the Arduino IDE software, including input and output programs, are run on the Arduino UNO itself as a platform. A notification in the form of a brief message indicating whether the lights are ON or OFF is generated by the SIM module 900A program, which acts as a message data processing medium. A light controller built from an electric magnetic plate, the relay module controls lighting that operates whether or not there is an electric current flowing.

11.6 PIR SENSOR AND RELAY MODULE

PIR sensors seem to be more confounded comparing a large number of different sensors (components such as photocells, slant switches, and FSRs) since there are different factors that influence the sensors' info and result. To start clarifying how a fundamental sensor works. Each of the PIR sensor's apertures is constructed of a special substance that is sensitive to IR light, and the PIR sensor itself has two openings in it. When the sensor is deactivated, both apertures recognize the same amount of IR radiation, which is the total amount of radiation that has originated from the room, the separators, or outside it. At the point when a warm body like a human or creature comes nearby the sensor, it first blocks one portion of the PIR sensor, which causes a positive differential change between the two parts. The converse happens whenever the hot body departs the detecting region, which results in the sensor producing a negative differential change. These alternating beats are what make the pattern recognizable. Figure 11.4 shows the PIR sensor description.

11.6.1 Working of the System

The very basic steps for how this system works are explained here. The PIR sensor initially fails to recognize any individuals while there is no human development, as well as its OUT pin, stays low. As the lone person approaches the place, an IR change in the place is noticed in the PIR sensor. Due to this, the PIR sensor returns a HIGH result. Since Digital Pin 8 of Arduino is connected to the PIR sensor's Data OUT, whenever it becomes HIGH, Arduino will start the hand-off by making the transfer pin LOW (as the hand-off module is a functioning LOW module). The light will now be ON. As long as there is construction in front of the sensor, the light remains ON.

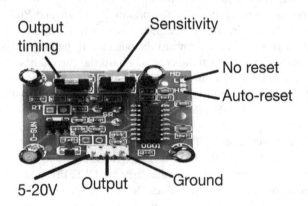

FIGURE 11.4 PIR sensor.

Automatic Room Light Controller Using Arduino and PIR Sensor

If the individual does turn it OFF, radiation IR intensity will be constant (none of that will change), and as a result, output data of the PIR sensor will be dropped. The room light will then be turned off and the Arduino will wind down the transfer by setting the hand-off pin HIGH.

11.6.2 THE CIRCUIT DIAGRAM

The Yield pin becomes HIGH when a PIR sensor senses human activity, applying the triggering voltage to the semiconductor's base, turning it ON, and allowing current to start flowing through the loop. A much larger current (220 V AC) can stream as a result of the electromagnetic field created by the loop in the relay, turning ON the bulb. By installing a PIR sensor, you can extend or shorten the time that the light is ON. Due to this, the PIR sensor returns a HIGH result. Since Arduino's Digital Pin 8 is linked to the PIR sensor's Data OUT, whenever that pin is HIGH, the hand-off will be implemented by Arduino by setting the hand-off pin to LOW (as the hand-off module is a functioning LOW module). The light will now be ON. As long as there is construction in front of the sensor, the light remains ON. The IR radiation will stabilize if the person falls asleep or leaves the room (there won't be a change), which will cause the PIR sensor's Data Out to drop. As a result, the Arduino will wind down the transfer (making the hand-off pin HIGH) and turn off the room light. Ardiuno working with PIR is shown in Figure 11.5.

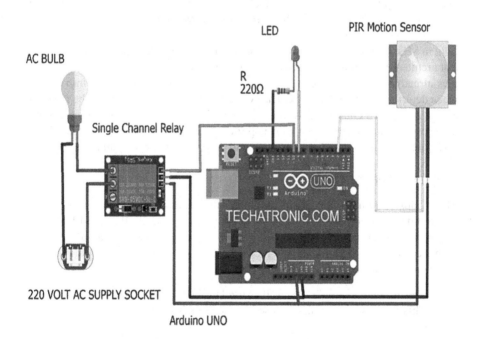

FIGURE 11.5 Ardiuno working with PIR.

11.7 RESULTS

The suggested methodology has been put into practice, and the Thingspeak cloud channel is being used to monitor real-time data. For office room automation, As soon as the preprogrammed set value has passed, all appliances turn off automatically (in this case, 3 minutes) in the event that the PIR sensors do not pick up any signs of occupancy, and the lights, fan, and air conditioners are turned ON in response if there is any indication of occupancy in any part of the space. The constructed system transmits real-time gathered data to the Thingspeak cloud channel at regular sample intervals.

11.8 CONCLUSION

The suggested framework leads us to infer that there is a procedure for controlling room lighting by means of numerous gadgets. A great deal of energy is wasted just in our day-to-day lives these days. With the help of this system, energy waste may be prevented, which contributes to significant force savings. The absolute viable cost of the framework is remarkably low. It is crucial to have adequate lighting in order to create ideal working conditions. The optimization of energy use inside buildings is another significant factor, primarily because of the current energy-related regulations. To achieve this, it is required to combine advancements in building structural designs that maximize natural light with suitable control strategies. The primary goals of this chapter are to design a system, validate it inside a conference room, and allow for preserving visual comfort conditions. More particular, this system has undergone rigorous testing inside a bioclimatic building's meeting room. These early findings indicate that the suggested structure is now considered to be successful in maintaining the required visual comfort levels while also reducing the use of artificial light in SHs. Future work will involve conducting actual tests with the suggested control architecture throughout various seasons and comparing it to other control strategies already in use. Other glare evaluation indexes will also be taken into account when determining the DGI index. The system's actual configuration will be altered to enable distributed control across the room's various zones. Last but not least, it will be incorporated into a multi objective optimization problem that will enable efficient user comfort control for thermal, visual, and air quality in SHs of smart cities.

11.9 FUTURE SCOPE

The proposed framework bases its judgments on whether or not humans are present, but in order for it to perform better, we can also communicate through LDR (Light Subordinate Resistor) in addition to the PIR sensor. This system can also interact with the Bluetooth module, allowing for simple one-tick control of the entire system from a mobile device.

1. This framework might very well be used in classrooms, schools, and so forth.
2. This framework can also be used in the home's bathrooms, stairways, and other areas.

These are the main traits of the developed unit that distinguish it from rival intelligent automation devices. The usage of technologies based on wireless communication, such as RF communication and cloud networking, and Wi-Fi is employed in the connection between pieces of equipment as well as that are imposed by cable communication and, as a result, saves money. Due to the utilization of a cloud network, automation systems can be observed from any location in the world. Additionally, by utilizing technologies like WiGig, Zigbee, and LoRa, this technology can be created more efficiently and at a lower cost. The design and implementation of an activity-based color mode LED lighting and dimming system is another accomplishment of this system. This is significant for both homes and many other fields that use artificial lighting, such as agriculture. Because of the comforting effects of color, it lowers energy usage. There are numerous ways to increase the productivity of the agricultural and industrial sectors. As a result, this has commercial value for industries like farming, tourism, lodging, and residential developments. Utilizing contemporary technology and experimentally demonstrating the accuracy of the provided colors can improve the output quality of an LED lighting system. The efficiency of the light dimming system can be increased by adding a few light sensors to strategic locations like doors and windows. In order to satisfy everyone, the ceiling fan and air conditioning control system is made to be employed in areas with large crowds in accordance with their preferences. As it operates in accordance with the suggestions of all users in one location, this system would be successful. As the ceiling fan control system automates in accordance with the user's skin wetness, this is crucial for the health of the skin. The amount of skin moisture was measured in this study using a different sensor. However, it is advised that the ideal course of action would be to create a computerized system that is able to collect information from either a smart ring, smartwatch, or fitness band, each of which contains a skin moisture detection sensor. Depending on the client's skin state, it can provide the automation system with correct data directly. It has been suggested that an alternative could be provided for people who would rather not have been using an intelligent watch, intelligent band, as well as other physical equipment if tests are conducted out to build a bed sheet or blanket in such a manner that it is able to collect data on the skin moisture. In this study, a QR code is used in conjunction with the system of the mobile application and automated controls. The QR code may be added on the back of the seats when employing this automation system in locations like lecture halls so that users can readily access it. By utilizing technology like NFC and RFID, this can be changed. This idea can also be further modified by making the appropriate adjustments.

REFERENCES

1. https://www.edgefx.in/automatic-room-light-controller-for-home-automation-applications
2. https://www.projectsof8051.com/automatic-room-light-controller-with-visitor-counter
3. B. Bharmal, A. Shahapurkar, and A. Aswalkar, "Automatic home lighting solutions using human detection, sunlight intensity and room temperature," International Research Journal of Engineering and Technology (IRJET) e-ISSN: 2395-0056, vol. 04, no. 06 I, pp. 691–696, June –2017.
4. https://www.electronicshub.org/automatic-room-lighting-system-using-microcontroller
5. https://circuitdigest.com/electronic-circuits/automatic-room-lights-using-pir-sensor-and-relay

6. J. Byun, S. Park, B. kang, and I. Hong, "Design and implementation of an intelligent energy saving system based on standby power reduction for a future zero-energy home environment," IEEE Transactions on Consumer Electronics, vol. 59, no. 3, pp. 507–514, 2013.
7. J. Feng, and Y. Yang, "Design and implementation of lighting control system for smart rooms," 2017 2nd IEEE International Conference on Computational Intelligence and Applications (ICCIA), 2017, pp. 476–481, https://doi.org/10.1109/CIAPP.2017.8167263
8. S. A. Sudiro, B. A. Wardijono, and D. A. Rohman, "Automatic controller component development using FPGA device," 2016 International Conference on Informatics and Computing (ICIC), 2016, pp. 122–127, https://doi.org/10.1109/IAC.2016.7905701
9. J. M. Rodríguez, M. Castilla, J. D. Álvarez, F. Rodríguez, and M. Berenguel, "A fuzzy controller for visual comfort inside a meeting-room," 2015 23rd Mediterranean Conference on Control and Automation (MED), 2015, pp. 999–1006, https://doi.org/10.1109/MED.2015.7158888
10. P. Valíček, T. Novák, J. Vaňuš, K. Sokanský, and R. Martinek, "Measurement of illuminance of interior lighting system automatically dimmed to the constant level depending on daylight," 2016 IEEE 16th International Conference on Environment and Electrical Engineering (EEEIC), 2016, pp. 1–5, https://doi.org/10.1109/EEEIC.2016.7555604
11. R. Leandros, W. J. W. Saputra, D. F. Murad, and D. Atmaja, "SMS notification on-off room lights with body detection using microcontrollers," 2020 International Conference on Information Management and Technology (ICIMTech), 2020, pp. 288–293, https://doi.org/10.1109/ICIMTech50083.2020.9211169
12. M. Singh, and S. L. Shimi, "Implementation of room automation with cloud-based monitoring system," 2018 2nd International Conference on Inventive Systems and Control (ICISC), 2018, pp. 813–817, https://doi.org/10.1109/ICISC.2018.8398911
13. S. B. M. S. S. Gunarathne, and S. R. D. Kalingamudali, "Smart automation system for controlling various appliances using a mobile device," 2019 IEEE International Conference on Industrial Technology (ICIT), 2019, pp. 1585–1590, https://doi.org/10.1109/ICIT.2019.8755104
14. R. K. Kodali, and A. C. Sabu, "Aqua monitoring system using AWS," 2022 International Conference on Computer Communication and Informatics (ICCCI), 2022, pp. 1–5, https://doi.org/10.1109/ICCCI54379.2022.9740798
15. S. Premalatha, T. S. Kumar, K. Srividya, B. Rajapandian, and H. Maadhavan, "Multi-way switching system using IoT," 2021 4th International Conference on Computing and Communications Technologies (ICCCT), 2021, pp. 215–218, https://doi.org/10.1109/ICCCT53315.2021.9711792
16. Vibhuti, and S. L. Shimi, "Implementation of smart class room using WAGO PLC," 2018 2nd International Conference on Inventive Systems and Control (ICISC), 2018, pp. 807–812, https://doi.org/10.1109/ICISC.2018.8398910
17. L. S. Karumuri, and A. Yarlagadda, "Smart rooms," 2020 IEEE 17th India Council International Conference (INDICON), 2020, pp. 1–4, https://doi.org/10.1109/INDICON49873.2020.9342283
18. J. Gubbi, R. Buyya, S. Marusic, and M. Palaniswami, "Internet of Things (IoT): A vision, architectural elements, and future directions," Future Generation Computation Systems, vol. 29, no. 7, pp. 1645–1660, 2013.
19. M. Inoue, T. Higuma, Y. Ito, N. Kushiro, and H. Kubota, "Network architecture for home energy management system," IEEE Transaction on Consumer Electronics, vol. 49, no. 3, pp. 606–613, 2003.
20. P. Hu, "A system architecture for software-defined industrial internet of things," 2015 IEEE International Conference Ubiquitous Wireless Broadband, ICUWB 2015, 2015.
21. Y. Jie, J. Y. Pei, L. Jun, G. Yun, and X. Wei, "Smart home system based on IoT technologies," Proceedings - 2013 International Conference on Computational and Information Sciences, 2013, pp. 1789–1791.

22. S. B. Mat, M. Y. Sulamiman, and K. Sopian, "Energy and buildings," Center for Science and Environment, vol. 62, no. 13, 2013.
23. I. I. Attia, and H. Ashour, "Energy saving through smart home," Online Journal on Power Energy Engineering, vol. 2, no. 3, pp. 223–227, 2011.
24. A. B. Jaffe, and R. N. Stavins, "The energy-efficiency gap what does it mean?" Energy Policy, vol. 22, no. 10, pp. 804–810, 1994.
25. L. Pérez-Lombard, J. Ortiz, and C. Pout, "A review on buildings energy consumption information," Energy Buildings, vol. 40, no. 3, pp. 394–398, 2008.
26. N. Aste, M. Manfren, and G. Marenzi, "Building automation and control systems and performance optimization: A framework for analysis," Renewable Sustainable Energy Reviews, vol. 75, no. September 2015, pp. 313–330, 2017.
27. S. P. S. G. N. K. Suryadevara, S. C. Mukhopadhyay, and S. D. T. Kelly, "WSN-based smart sensors and actuator for power management in intelligent buildings," IEEE/ASME Transactions Mechatronics, vol. 20, no. 2, pp. 564–571, 2014.
28. J. King, and C. Perry, "Smart buildings: Using smart technology to save energy in existing buildings," American Council for an Energy-Efficient Economy, February, pp. 1–46, 2017.
29. A. Al-fuqaha, S. Member, M. Guizani, M. Mohammadi, and S. Member, "Internet of Things: A survey on enabling," IEEE Communications Surveys & Tutorials, vol. 17, no. 4, pp. 2347–2376, 2015.
30. NASSCOM, "The future of internet India," Proceeding of the Annual meeting (American Society of International Law), vol. 101, July, 2016.
31. S. Li, L. Da Xu, and S. Zhao, "The internet of things: A survey," Information System Frontiers, vol. 17, no. 2, pp. 243–259, 2015.
32. P. Sindhuja, and M. S. Balamurugan, "Smart power monitoring and control system through Internet of Things using cloud data storage," Indian Journal Science Technology, vol. 8, no. 19, pp. 1–8, 2015.
33. N. Nesa, and I. Banerjee, "IoT-based sensor data fusion for occupancy sensing using Dempster Shafer evidence theory for smart buildings," IEEE Internet Things Journal, vol. 4662, no. 5, pp. 1563–1570, 2017.
34. S. D. T. Kelly, N. K. Suryadevara, and S. C. Mukhopadhyay, "Towards the implementation of IoT for environmental condition monitoring in homes," IEEE Sensors Journal, vol. 13, no. 10, pp. 3846–3853, 2013.
35. M. Cabras, V. Pilloni, and L. Atzori, "A novel Smart Home Energy Management system: Cooperative neighbourhood and adaptive renewable energy usage," IEEE International Conference Communication, London, UK,, 2015, pp. 716–721.
36. P. Singh, and S. Saikia, "Arduino-based smart irrigation using water flow sensor, soil moisture sensor, temperature sensor and ESP8266 WiFi module," IEEE Reg. 10 Humanity Technology Conference 2016, Agra, India, R10-HTC 2016 - Proceedings, 2017.
37. S. Saha, "Data centre temperature monitoring with esp8266 based wireless sensor network and cloud based dashboard with real time alert system," IEEE International Conference on Devices for Integrated Circuits (DevIC), Kalyani, India, 2017, pp. 23–24.
38. A. Namburu, and S. S. Barpanda (Eds.). (2020). Recent Advances in Computer Based Systems, Processes and Applications: Proceedings of Recent Advances in Computer based Systems, Processes and Applications (NCRACSPA-2019), October 21–22, 2019 (1st ed.). CRC Press. https://doi.org/10.1201/9781003043980
39. N. Bit t erman, and D. Shach-Pinsly, "Smart home – A challenge for architect s and designers," Architectural Science Review, vol. 58, no. 3, pp. 266–274, 2015. https://doi.org/10.1080/00038628.2015.1034649
40. S. Ghosh, "Smart homes: Architectural and engineering design imperatives for smart city building codes," 2018 Technologies for Smart-City Energy Security and Power (ICSESP), Bhubaneswar, India, 2018, pp. 1–4, https://doi.org/10.1109/ICSESP.2018.8376676

12 Role of IoT in Supply-Chain Management Processes

Usha Yadav and Sheetal Soni

12.1 INTRODUCTION

In order to compete in the modern business management environment, individual firms must participate actively in the larger supply chain (SC), which is a network of several enterprises and partnerships [1]. Supply-chain management (SCM) and logistics have seen notable transformations in past few years in upcoming smart cities [2]. Competitive pressure has fuelled the growing interest in SCM and logistics, which has eventually elevated it to become a crucial component of business operations and strategy [3]. The fourth industrial revolution's new business prospects and technology developments provide the manufacturing sector and its SCs opportunities for performance improvement that were previously unthinkable. As a result, supply chains operate in a constantly shifting environment and are open to a wide range of hazards on all fronts. This environment is a dynamic terrain that is influenced by numerous elements [4]. Many supply networks are exposed to several global risks because they span large geographic areas [5]. Customers are becoming more and pickier in terms of product customization, cost, and service quality [6]. In addition, the external environment is quite dynamic because of economic, social, and ecological elements including the cost of energy, the pricing and availability of raw materials, and currency exchange rates as well as the unavoidable natural factors such as extreme weather conditions, earthquakes, and tsunamis [4]. Nevertheless, advances in instrumentation, connectivity, and intelligence can produce the reliable, secure, and sustainable supply chains that today's enterprises require [5]. The goal of current SCM practice has been to achieve what is refer to as dynamic flexibility, which enables businesses to adapt to specific changes in demand and technological advancements, but only within the predetermined framework of their current SC designs [6]. However, if SC architecture is to tackle the challenges of a dynamic economic environment, structural flexibility that embraces flexible alternatives is required. However, it will take some time for the Internet of Things (IoT) to fully take hold in a world where the majority of businesses are still utilizing a combination of out-of-date legacy systems, new technologies, and connected and disconnected assets. The IoT, the primary player in Industry 4.0, provides advantages for the whole logistics value chain by enhancing SC connectivity and visibility [7]. Therefore, the goal of this study is to examine how IoT technologies enable SC transparency and connectivity as well as their effects on SC performance.

12.2 INCREASING COMPLEXITIES OF SUPPLY-CHAIN PROCESSES

Supply chains are now more vulnerable to natural disasters and interruptions as a result of their increased globalization and connectivity. Today, a crisis in one far-off nation can spread rapidly to global economy. No company is safe as supply chains become more interconnected. The first Chief Supply-Chain Officer Study, conducted by IBM in 2008, was focused on personal, in-depth interviews with the top-notch SC executives of companies, of which there are over 400 spread across 25 nations in Asia Pacific, Western Europe, and North America. These executives are in charge of supply chains that diverse sectors, including retail, manufacturing, food and beverage, medicines, electronics, and telecom [5]. The study emphasizes the five main challenges that contemporary SCM is facing. The challenges are shown in Figure 12.1.

Cost-cutting is the top priority for SC managers, far above business growth and new product/service development. Commodity price rises, unexpected loan restrictions, and abrupt rising wages in previously lower-wage markets are all appearing more frequently. And because of this, businesses continually work to improve the efficiency of SC activities in an effort to cut costs. The SC managers are also struggling with SC visibility. Contrarily, despite the fact that there is more information available, less of it is effectively managed, assessed, and made available to those who need it. The study found that many executives claimed their companies were either simply too preoccupied to exchange real sales data or did not value collegial

FIGURE 12.1 Challenges in supply-chain management.

decision-making. Risk management is also another topic that worries SC managers. Risk has increased due to globalization and the wider supply chain's interconnection, but they have also made it more challenging to manage. Managers point to a lack of consistent risk management procedures, an absence of data, and inappropriate technologies as the main roadblocks. Customer involvement in the various stages of product design is the main issue SC managers are dealing with. The study states that two out of every three businesses have trouble in correctly identifying customer needs. Although it's clear that businesses need to communicate with their customers, they frequently place more of an emphasis on their suppliers. Globalization is among the top SC challenges, which should not be a shock given the increasing interconnectedness of economies around the world. Many businesses are having trouble with international procurement because of inconsistent delivery, lengthy lead times, and substandard quality.

The study entails that the number of enterprises involved in supply chains has increased along with their geographic reach. As indicated in Figure 12.2, majority of managers indicated they anticipate a rise in the number of joint interactions with outside parties [8]. Additionally, wider spectrums of tasks are being subcontracted by 2010. Business companies must proactively incorporate intelligence into their decision-making and management systems in order to address these challenges. Companies need to be aware of the latest technologies that are currently influencing SC operations if they want to maintain a competitive advantage in today's rapidly evolving environment. Using digital tools to monitor and track things from their source of basic materials to their customer can provide complete SC visibility. The objective is to keep the data format uniform across all vendors and chain components. Digital devices can then be connected to the data using technology such as blockchain technology, a common platform where many participants in the supply chain can access IoT, radio frequency identification (RFID), chain input and verification, and artificial intelligence. The inherent SC challenges can, to a certain extent, be overcome with the aid of the IoT.

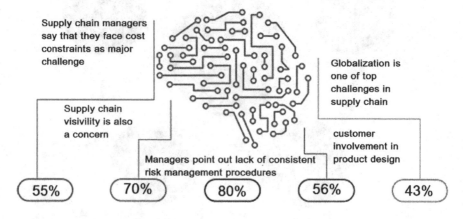

FIGURE 12.2 Five major challenges faced by supply-chain leaders.

12.3 EVOLUTION IN SUPPLY-CHAIN ENVIRONMENT

Supply chains evolve over time in terms of their size, shape, configuration, and techniques for coordinating, controlling, and managing them. New supply chains may emerge for a variety of reasons, such as in response to technological innovation [9]. It is suggested that the structure, scale, and kind of trade are ultimately determined by fundamental economic factors rather than SC features [10]. However, other factors may also have a significant impact over time on how a supply chain is configured, run, and coordinated. Regulatory frameworks [11], sustainability agendas, political factors [12], and strategic decisions [13] have an impact on the structure and configuration of supply chains in addition to economic and technological drivers. Global sourcing techniques have drastically altered how supply networks are set up [14]. In order to follow a manufacturing and/or marketing plan and better serve their markets, organizations have strategically re-engineered their networks. Thus, a variety of economic, technological, environmental, and strategic factors could potentially have an impact on the SC participants, the locations of value-adding activities, the coordination and management of these activities, as well as the growth and development of supply networks.

The fourth industrial revolution is being represented by the new system of applications and services such as the IoT, cyber-physical systems (CPS), blockchains, big data, and virtualization. According to numerous studies, the fourth industrial revolution, often known as Industry 4.0, is the fusion of existing information and communication technologies for the creation of goods and services [15]. The most prominent study subjects among practitioners and scholars are the IoT and its advantages and difficulties. Increased SC visibility would assist businesses in streamlining operations and reducing complexity by reducing inaccuracies as SCs become more complex at a rapid rate [16]. SCM's importance will grow in the upcoming years as a result of increased globalization, customer focus, and improvements in information technology (IT) [17]. Implementation of various technologies, including IoT, has largely favorable effects on a variety of sectors, including the production of energy, manufacturing, warehousing, healthcare, chemical and pharmaceutical, and food and beverage industries. Continuous technological developments, such as augmented reality, direct digital production, warehouse automation, and 3D printing, to name a few, open up a whole new world of possibilities from an SC perspective. In addition to improving the productivity and adaptability of production and distribution operations, these new technologies also change the dynamic between the various SC nodes, with a focus on the consumer [18].

The goal of the intelligent SC is to interconnect the processes in the many partners of an SC to create an intelligent connected system using IoT, blockchain, etc., by utilizing cutting-edge technologies, particularly rising information and communication technology (ICT) [19]. ICT and other upcoming technologies from smart manufacturing have the potential to significantly improve performance of SCs. Innovative technologies, particularly cutting-edge ICT, are what are driving the fourth industrial revolution, often known as Industry 4.0. Almost all phases of the manufacturing system are impacted by sophisticated ICT [20]. The study of product and need information flows and their corresponding coordination receives a lot of attention in modern research on SCM [21]. Since all SCs are fundamentally driven by demand, data on end consumer demand is frequently regarded as the most crucial data in SC networks [19]. Availability, revenue, pricing, transportation, region, volume, performance, and technical data are examples of more common information that is produced [21]. In the age of big data, new technologies are necessary to acquire, store, distribute, and analyze massive volumes of fast-moving, complex, and diverse data. The IoT is a technical advancement in communications and computing that will enable "anytime, anywhere, any media, everything" interactions [22]. As a result, IoT makes it possible to connect virtual and physical elements, opening up a whole new range of services and applications.

12.4 INTERNET OF THINGS

The growth in the era of IT has been crucial in order to improve the planning, execution and control of the flows and storage of commodities, data and services from the place of initiation to the place of consumption. The speed of change brought on by new technology has revolutionized how companies produce and distribute value to customers. For instance, Industry 4.0's introduction has been demonstrated to be a crucial success element for delivering numerous business benefits, such as the optimization of value chain activities in industries and day-to-day operations and processes [23].

The German Economic Development Agency [24] invented the phrase "Industry 4.0" to describe the appearance, development, and convergence of numerous technologies that enable nearly real-time interaction between the physical and digital worlds. As a result, technology has been and still is a crucial facilitator for efficient SCM. Through bettering communication, data acquisition, and data transmission, it plays a crucial part in coordinating the actors, enabling efficient decision-making and boosting SC performance. One of the most recent IT advancements in SCM is the IoT, which can offer more precise data for more efficient decision-making.

12.5 ROLE OF IoT IN SUSTAINABLE SUPPLY-CHAIN MANAGEMENT

There are numerous IoT definitions in the literature. Some of them are defined as "It is a technical idea where several connected devices are able to turn on and off the internet in order to leverage software and automated procedures for smart applications." RFID tags can establish communication and transmit the identifying data when connected over a network. [25].

The effectiveness and efficiency of delivery, ongoing monitoring, and close partnerships between partners are only a few of the problems that afflict the SC. The IoT creates a network of physically connected things that the stakeholders can continuously monitor and detect. As a result, increased data transparency and visibility are made possible by the use of this technology, such as temperature monitoring, locations, humidity, pressure, light exposure, and broken seals. IoT also reduces costs and waste. IoT is technically based on fundamental technologies like RFID and GPS that are then connected to the internet [3].

IoT applications have been widely implemented in the present scenario to find, recognize, track, and monitor goods and services in the SC [16]. Additionally, in this instance, the technology has been applied in other disciplines, particularly in SCs for the agricultural and medical industries. The foremost crucial obstacle in executing the IoT infrastructure in the area of SCM is security concerns, a lack of standards, interoperability issues, and software and hardware restrictions. Although RFID technology has been a key IoT enabler for many years, there are still problems with pricing, technological adoption, complexity, scale implementation, and data management [22].

The data management difficulties that arise as a result of smart objects sharing a vast quantity of information and connecting resources that can be accessed through a web interface are addressed in the semantic-oriented component. Devices and intelligent items, such as sensors, actuators, and internet-connected RFID chips. Since the advent of wireless technology, the IoT has gained greater attention [10], and the SCM community has also shown interest in it [16, 26]. IoT has made significant contributions to industrial automation and enabled the fusion and integration of networks for plant control, corporate information management, and RFID networks for logistics management.

In addition, IoT has given businesses ways to boost operational effectiveness, guarantee convenience in their operations, and maintain a competitive edge. Businesses have the ability to employ IoT to streamline information flows, provide significant SC efficiency improvements, and enable communication and integration between and within organizations [27]. For instance, the international fashion retailer Zara has achieved success in maintaining high planning flexibility, reliable replenishment solutions, shorter lead times, and fewer product changes based on IoT [28].

12.6 DETAILED LITERATURE ON SUPPLY-CHAIN PROCESS AND IoT

SCM benefits from the IoT's cost-saving, inventory accuracy, and product tracking features, among others. However, the influence of IoT on various SC activities is unknown [4]. The process-centric view of the SC is one paradigm for comprehending SCs. An illustrative framework for this is the Supply Chain Operation Reference Model (SCOR) model, which breaks down the processes of SCs under Plan, Source, Make, Deliver, and Return. Therefore, the purpose of this research is to identify the influence of IoT on SCM by conducting a comprehensive literature review based on the affected SC processes. Given that the Plan process is intertwined with all other SCOR processes, the present literature analysis concentrates on the remaining SCOR processes.

12.6.1 SOURCE

An upstream component of the SC is sourcing, which is the process of carefully deciding which products and services a firm requires to perform its business. In addition to choosing a seller, negotiating a contract, and assessing your suppliers' long-term performance, sourcing also refers to the act of purchasing goods. The ability to quickly accept and react to changes in the business environment is made possible by virtual SCs. In response to industry constraints and opportunities provided by today's accessible new technology, SCs are becoming more virtualized [29]. The ability to change organizational structures allows it to design solutions to meet changing client needs. This is made possible by the virtual SC [30]. In essence, virtual SCs disentangle management tasks from the actual movement of goods from source to destination. Instead of being directly observed by real objects, virtual objects are used to regulate and coordinate SC activities [29]. Also discussed is the Track & Trace Analytics framework, which may be expanded to provide the capability to monitor product condition as it moves through SCs. This project is called "Building Radio frequency Identification for the Global Environment" and is partially financed by the European Union. The report goes into detail about taking into account various approaches to fusing sensor data with RFID/EPC analyses, through either sensor-enabled labels or by using virtual integration in which sensor information from environmental sensors is directly linked with the identified positions of objects at various times [31]. Another study shed light on the IoT's guiding principle and its consequences for big data analysis on the operating efficiency of SCs, especially in relation to the kinetics of operational coordination and SC optimization by utilizing big data acquired from interconnected smart products, and the governance framework of big data disclosure [32].

12.6.2 Make

The foundation of product manufacturing is the ability to produce a high-quality good within the required time frame and budget. All effective SCs establish and regularly review the criteria they use to decide on expenses, values, quality, and turnaround time at each stage of the production process, from supplies and labor to storage and transportation. The four stages of manufacturing's historical development are known as Industries 1.0 through 4.0. Each of the phases was a significant paradigm shift in manufacturing. With the use of steam and water power, industrial production was first introduced in Industry 1.0. Industry 2.0 was characterized as mass manufacturing due to the specialization of labor facilitated by electrical energy. In order to further automate production, Industry 3.0 introduced electronics, IT, and control mechanisms to the shop floor. With the aid of IoT, Industry 4.0 promises an unparalleled radical shift that will have significant effects on production and its supply network [4]. Cloud computing makes use of superpowerful computational tools to solve problems and efficiently facilitates seamless connections between people and objects to resolve the difficult matters with decision-making [33]. A framework for real-time accessibility and transparency is recommended in accordance with the analysis of the increased needs for shop floor management encountered by current discrete manufacturing businesses [34].

12.6.3 Deliver

Logistics are "part of the supply chain process that plans, implements and controls the efficient, effective forward and reverse flow and storage of goods, services and related information between the point of origin and the point of consumption in order to meet customer requirements," according to the Council of Supply-Chain Management Professionals.

In one of the research, an approach for future-generation warehouses was suggested, wherein an RFID-based inventory management system was proposed that includes automated storage and retrieval mechanism that uses RFID technology and can communicate with an RFID-based system for inventory management without the need for human interaction [35]. Another study backed the use of RFID to enhance the flexibility of decentralized warehouse management in a changing environment [29]. Table 12.1 explains the deployment of RFID-based system and the development of a prototype for improvement in the effectiveness of the SC. The majority of the studies highlighted inventory administration and warehousing.

12.6.4 RETURN

From the perspective of operations management, the reverse logistic phenomenon has drawn a lot of attention from both the business and academic worlds. The kinetics of management of the inventories are impacted by the reverse flow of goods entering the chain. This, in return, has an impact on the dynamics of orders submitted to suppliers and, as a result, has an effect on how well the SC performs in terms of order and inventory fluctuation [36]. Table 12.1 highlights the implementation of IoT to make closed-loop SC feasible for businesses. Intelligent goods that surround memory, sensing, data dealing out, reasoning and communication facility, and their crucial function in product lifecycle management were also introduced in the research. More integrated software systems, such as intelligent embedded systems that give real-time data to information management systems at a higher level, are on the rise. [37]. There is a need for a reverse management information system that can both vertically and horizontally record product qualities and life cycles, assisting businesses in improving decision-making and operational efficiency [38]. IoT in the home also has enormous potential outside of existing business models because downward data provides insight into how consumers consume resources, increasing the opportunity to develop novel offers, conserving the environment, and aiding forward and reverse SCs that can interact to use in perspective [39]. Consumers control the big data from IoT, and conversations for access to such data lay the groundwork for marketplaces for both the raw data and apps that modify it for the advantage of customers and businesses. Such information is important to businesses if they can exchange rights to it with customers in exchange for acceptable adherence to security, privacy, and reliability [40].

12.7 RFID FOR INNOVATIVE SUPPLY-CHAIN MANAGEMENT

Numerous features like unique product identification, simple connectivity, and real-time information availability, RFID technologies have a variety of benefits for the SC [26]. RFIDs are specifically tools for obtaining data about a certain product. They are transferred to the system via specialized readers and include a lot of data. Although the use of RFID in SCs is not new, numerous experts are still researching

TABLE 12.1
Literature Review of Supply-Chain Process and IoT

References	SCM Processes	IoT Interface	Supply-Chain Improvement
[29]	Source	The architecture of information systems to support virtual supply chains	Enhancement of supply-chain visibility better quality, and shorter turnaround times
[30]		Modular blocks to further integrate at the larger level of supply chain	Flexibility to adjust organizational structures so that it may tailor solutions to keep up with shifting customer needs
[31]		Developed Track & Trace Analytics framework to monitor condition of products	Improved supply-chain efficiency in a cost-effective manner, quality assurance, and stock control
[41]		Suggested tailoring and platform strategy for choosing production, suppliers, and assortments	Postponement, product differentiations, benefits for choosing the desirable strategy to maximize the consumer value
[32]		Big data analytics' effects on the operational effectiveness of the supply chain and the implications of IoT, governing framework for accessing big data	Utilizing big data from connected smart products for operational coordination and supply-chain optimization
[42]		Suggested a triadic model based on the asset-process-performance paradigm that encompasses online retailers and service providers for product delivery	Criteria for e-retailers selling innovative products to choose product delivery services, and its impacts on achieving the customer satisfaction
[33]	Make	The data sources of next-generation manufacturing companies can be efficiently enabled by growing IoT infrastructure in the dynamic environment	Cloud computing makes use of superpowerful computational tools to solve problems and efficiently facilitates flawless connections between people and objects
[34]		A framework for real-time accessibility and transparency is suggested	Real-time data collection and analysis for contemporary manufacturing would save lead times and costs
[43]		Deployment of RFID to production lines for controlling and monitoring overall production in an augmented reality environment	RFID, infrared-enhanced computer vision, and inertial sensor application techniques are aiming to provide just-in-time information rendering and intuitive information navigation
[44]		Smart objects embedded production and quality management functions	Assisting with production and quality management decisions

(Continued)

TABLE 12.1 (Continued)
Literature Review of Supply-Chain Process and IoT

References	SCM Processes	IoT Interface	Supply-Chain Improvement
[35]	Delivery	An automated storage and retrieval mechanism that uses RFID technology can communicate with an RFID-based system for inventory management without the need for human interaction	Maximize material handling operations' efficiency and cut operational costs
[29, 45]		An RFID-based storage capacity planning system helps decision support for storage assignment in a warehouse	Make order picking in a warehouse more productive; decentralized warehouse management
[46]		Suggested software architecture that utilized the benefits of RFID and Web 2.0 tools to create a prototype inventory management software	Assisted in identifying missing goods and lower inventory levels, and sent web-based notifications to inventory management
[31, 47]		Tracking of inventory and real-time information sharing	Instantaneous visibility; effective utilization of area and resources
[48]		Suggested sensor portfolios based on income centric values	Predicting shelf life, increasing sales, producing precisely, and eliminating verification costs
[37]	Return	Smart embedded systems for closed-loop life cycle	Transparent and uninterrupted connectivity
[38]		Deploying IoT in reverse logistics by connecting positive and negative logistics	Building closed-loop supply chain
[39]		Development of IoT system for real-time data capturing	Information about actual consumption

this subject [25]. In actuality, there are other areas that merit investigation, including inventory management, order management, warehouse management, and production scheduling. The general framework of RFID implemented in SCM is shown in Figure 12.3.

12.7.1 RFID IN INTEGRATION

A successful SC, especially a global SC, depends on effective integration—communication and information sharing. Operating your business would be similar to having the servers fill out four different forms only to send the kitchen crew an order if your SC isn't interconnected. It's absurd, and it won't help you much. Not integrating your SC is ridiculous. The more quickly suppliers, warehouses, and transportation companies can exchange information, the more efficiently your SC will move items. A stronger SC is one that is more tightly wound.

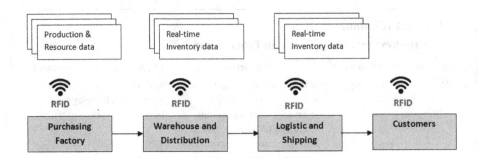

FIGURE 12.3 Usage of RFID in supply-chain management.

Data access and sharing are made simpler by RFID technology. RFID tags enable businesses to easily track down commodities that are in transit and collect information about them, which is subsequently sent to other SC participants. However, not every industry can fully integrate SC activities using RFID technology. Walmart, for instance, demanded in 2003 that all suppliers put RFID tags on their shipments within 18 months for its top 100 suppliers and by the end of 2006 for its other vendors. The initiative ultimately failed, mostly because it didn't appear to have any advantages. The extremely accurate scan rates promised by RFID technology didn't materialize since RFID can't penetrate metals and liquids. RFID technology is particularly suited for sectors including fashion and apparel, cosmetics, jewelry, healthcare, and pharmaceuticals where products don't obstruct RFID readability.

12.7.2 RFID in Operations

A company's SC can succeed or fail based on day-to-day operations. It's best summed up by the famous saying, "We are what we repeatedly do. Therefore, excellence is a habit rather than an act." SC managers that put a priority on outstanding day-to-day operations create an SC that is more than the sum of its parts and accomplishes two major goals at once: increasing revenue and reducing operational expenses. SC managers can assess how well they are performing on a daily basis by using performance data. Which procedures increase employee productivity? Where is a need for extra procedures? Where do untapped opportunities exist?

Companies can simply track the flow of goods through their warehouses with the aid of RFID scanners. Inventory management becomes considerably more effective as a result of businesses having access to real-time data on what is in the warehouse, what is being delivered to the warehouse, and what has been dispatched to customers. While RFID integrations with warehouse control systems offer the potential for drone capabilities in the future, RFID integrations with SCM systems enable businesses to obtain real-time information. How may this appear? In order to maintain track of goods, drones would essentially fly around warehouses and collect data from a few centimeters away. Although widespread implementation of this strategy is still a ways off, the potential of RFID technology in warehouses is alluring. One well-known business that has employed RFID tags to achieve operational efficiency is Zara. As a fast-fashion shop, Zara's whole business strategy depends on its capacity to provide affordable substitutes for trendy, more expensive items. One of the primary drawbacks of this business strategy is that trends come and go, yet suppliers may require many months' notice to build sufficient inventory.

The large international retailer employs RFID tags across the SC to keep track of its enormous inventory in real time. Zara encrypts clothing starting with the production process to enable end-to-end traceability. Inventory management can rapidly determine whether things are out of stock while also enabling Zara to produce more quickly in response to quickly moving merchandise. It is also important to note that Zara makes its own clothes.

12.7.3 RFID in Purchasing

If customers can't get the raw resources they require, they can't make the products they need. That is why purchasing is a crucial step in the SC process. Customers need the component pieces to make the finished product whether they are creating computers, fast-moving consumer packaged goods, or garments. Working with suppliers who have RFID technology allows customers to see raw materials and semi-finished products in stock in real time. RFID also makes it simple to rapidly verify that each shipment has what they bought, which cuts down on waiting time at the receiving stage and hastens the transition of goods to the production stage.

In the event that raw materials include flaws, RFID can offer some protection. Imagine, for instance, if one of your food suppliers reported contamination in a raw ingredient you use. If you and your supplier both use RFID technology, your data can be used to identify which completed goods that specific raw material was used in. Customers are safeguarded, and the process of getting paid by that provider for your business is accelerated.

12.7.4 RFID in Distribution

From the point of origin to the point of consumption, things are coordinated and moved through logistics. Whether it's the distribution of finished goods to warehouses, the distribution of individual items to customers, or the distribution of raw materials to manufacturers, effective logistics is essential. It's crucial for businesses that move goods to uphold quality control, achieve agreed-upon service standards, keep costs down, and more.

All of this is dependent on the accurate and timely transmission of information. It's crucial to be aware of the condition of your products if there is severe weather close to your suppliers. Have they left the danger zone or do you need to formulate a backup plan right away? Have there been any temperature changes to the shipping container that might have an impact on the integrity of delicate products like pharmaceuticals? Real-time information makes it feasible to make swift decisions about distribution via cargo.

12.8 IoT-BASED FUZZY LOGIC DECISION SYSTEM FOR REVERSED LOGISTICS

When crisp quantities are transformed into fuzzy ones, this process is known as fuzzification. Membership functions serve as an illustration of the translation into fuzzy values. The degree of each input's participation in a particular collection is represented graphically by a membership function. The membership function can be of any type depending upon the nature of the problem. The method of fuzzification may entail giving the provided crisp quantities membership values for distinct sets.

Rule-based reasoning is used by the fuzzy system to choose an output answer. It is a specific kind of logic that employs IF-THEN rules. To carry out the reasoning process, the Fuzzy Inference System assesses each rule [49]. The conceptual framework of fuzzy inference system is shown in Figure 12.4.

Defuzzification is the procedure used to transform fuzzy values into clear ones. The results produced are hazy and cannot be used as such. As a result, in order to process them further, fuzzy amounts must be transformed into crisp ones. Several techniques can be used to achieve this. However, Ref. [50] used the maximum membership principle method in their strategy. The output membership function (for all sets) is taken to be singleton in this simple way. The fuzzy logic-based technique begins by establishing a list of unprecedented occurrences to be included in the analysis, much like the traditional Trend Impact Analysis (TIA). A collection of related attributes is found for each such event. Then, the matching characteristics and range of fuzzy values for each attribute must be established.

Reverse logistics is the process of sending goods back up the SC from end users to either the producer or the retailer. Offering returns to your consumers is just smart business, whether the client is returning products they don't need, the product has reached the end of its life cycle, or the product is broken or defective. This is where reverse logistics comes into play. This procedure also applies when products need to be thrown away or recycled, and it covers the case in which the final user is in charge of the product's refurbishment, disposal, or even resale [51].

The method used to handle the return of items is determined by the kind of reverse logistics in question, the kind of business the organization does, and the industry in which it operates. For instance, although a coffee pod maker might provide clients with a mail-back recycling scheme for discarded pods, a clothing retailer would

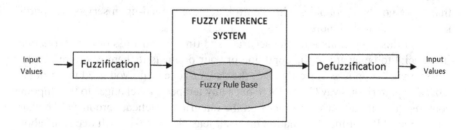

FIGURE 12.4 Conceptual framework of fuzzy inference system.

Role of IoT in Supply-Chain Management Processes

frequently deal with general customer returns. In order to recover their keg canisters and clean and refill them for onward distribution, breweries use reverse logistics. To ensure optimum efficiency and to take advantage of any incentives for optimizing its reverse-logistics processes to promote environmental sustainability, each business should have a documented process [52].

Consider a scenario in which companies want to take advantage of a product (vacuum cleaner) that is no longer in use by the customer or it needs to be disposed of. These products can be recycled by companies to gain financial benefits and maintain environmental sustainability. The major challenge is to predict the right time to reach the customer for product disposal or if the product is no longer in use. The fuzzification system can be used to predict the scenario in which customers would be willing to dispose of the product by using sensors installed inside the product to gather information about its maintenance history, health, utilization, and inactive mode.

12.8.1 Input 1: Frequency of Maintenance/Repair

The fuzzy set is represented by F, the elements of the sets are depicted by f_i and the membership function is represented by $\mu_F(f_i)$, and it is assumed to be triangular. The fuzzy set (f) is computed using Eq. (12.1).

The graphical representation of this is shown in Figure 12.5.

$$F = \sum_{i=1}^{n} \frac{\mu_F(f_i)}{(f_i)} \tag{12.1}$$

Here F = {L,M,H}, where L depicts the low value, M depicts the medium value, and H depicts the high value. The high value of frequency of maintenance/repair provides an indication that the customer is most likely to buy new product and would be willing to dispose off the old product.

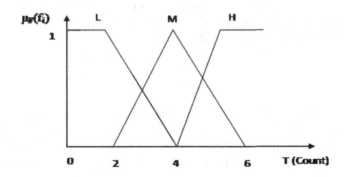

FIGURE 12.5 Fuzzy Set of $\mu_F(f_i)$.

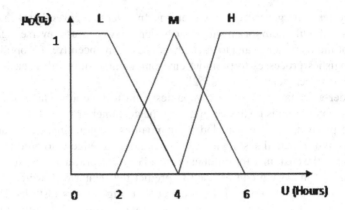

FIGURE 12.6 Fuzzy set of $\mu_U(u_i)$.

12.8.2 INPUT 2: TIME UTILIZATION OF PRODUCT

The fuzzy set is represented by (U), the elements of the sets are depicted by u_i and the membership function is represented by $\mu_U(u_i)$, and it is assumed to be triangular. The fuzzy set (U) is computed using Eq. (12.2).

The graphical representation of this is shown in Figure 12.6.

$$U = \sum_{i=1}^{n} \frac{\mu_U(u_i)}{(u_i)} \qquad (12.2)$$

U={L,M,H}, where L depicts the low value, M depicts the medium value, and H depicts the high value. It can be inferred from the time utilization of the product that the mentioned product plays an important role in the daily life of the customer.

12.8.3 INPUT 3: PRODUCT INACTIVE MODE

The fuzzy set is represented by (M), the elements of the sets are depicted by m_i and the membership function is represented by $\mu_M(m_i)$, and it is assumed to be triangular. The Fuzzy fuzzy set (M) is computed using Eq. (12.3).

The graphical representation of this is shown in Figure 12.7.

$$M = \sum_{i=1}^{n} \frac{\mu_M(m_i)}{(m_i)} \qquad (12.3)$$

M={VL, L, N, H, VH}, Where VL depicts the Very very low value, L depicts the low value, N depicts the normal value, H represents the High value, and VH represents the very high value. More is the value of the inactive mode. From that, it can be inferred that the customer is not in much need of the product and could be willing to dispose or resell.

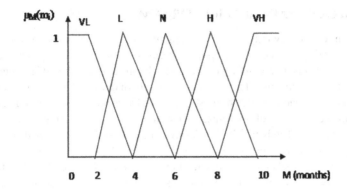

FIGURE 12.7 Fuzzy Set of $\mu_M(m_i)$.

12.8.4 OUTPUT: DISPOSE INTENSION

The fuzzy set is represented by *DI*, the elements of the sets are depicted by di_i and the membership function is represented by $\mu_{DI}(di_i)$, and it is assumed to be triangular. The fuzzy set (*DI*) is computed using Eq. (12.4). The graphical representation of this is shown in Figure 12.8.

$$DI = \sum_{i=1}^{n} \frac{\mu_{DI}(di_i)}{(di_i)} \qquad (12.4)$$

Here, DI={DN, NL, N, L, ML}, where DN depicts the "definitely not" value, NL depicts the not "likely value", N depicts the "neutral value", L represents the "likely value", and ML represents the "most likely value". If more is the inactive mode, then it can be inferred that the customer is not in much need of the product and could be willing to dispose of or resell it.

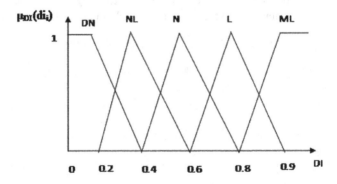

FIGURE 12.8 Fuzzy Set of $\mu_{DI}(di_i)$.

12.9 BLOCKCHAIN AND IoT FOR SCM

Blockchain technology may offer a permanent record of all transactions made along the SC, enabling businesses to spot problem areas and take appropriate action. Multinational firms that regularly deal with problems in the global SC have expressed interest in this possible use. The blockchain is a chain of linked blocks that serves as a distributed data structure. The basic structure of blockchain is shown in Figure 12.9. Blockchain keeps track of all transactions on a network by acting as a distributed database or a global ledger. The blocks that include the time-stamped transactions are recognized by their cryptographic hashes. The blocks are arranged in a linear chain called the "blockchain," with each block referencing the hash of the one before it. Every node in a network that maintains a blockchain executes and records the same transactions. Each node in the blockchain network has a copy of the blockchain. The transactions can be read by any node in the network.

The blockchain is a network-wide ledger that is safe, encrypted, unchangeable, and accessible. These features are necessary for players who must make important purchase selections because of product tracking [53, 54]. It has been used in numerous SC industries, including pharmaceuticals, agri-food, and marine shipping. The technology enables secure product tracing since the consensus techniques used by the distributed ledger ensure that the data within it remains unchangeable. IoT device use and blockchain SC applications are frequently mentioned together in literature [55]. When using only RFIDs and IoT infrastructure, there is a lack of security [56].

In order to validate and permit actions by physical devices, smart contracts can be used to optimise the entire SC, increasing the security of the IoT platform. Associating tracking tools and blockchain technology with smart contracts that enable payments to suppliers once they complete their tasks, such as delivering goods to a warehouse or meeting a predefined quality and quantity specification, is one way to use both in transportation and logistics [57]. In this scenario, RFID sensors are used to record information on the physical state of the items in the carrier's vehicle and the producer's warehouse. Data must be transferred from the RFID sensors placed adjacent to the items to the distributed ledger via the IoT infrastructure. Finally, blockchain ensures that data is recorded safely and forever and that players can see all transactions. The transferred data is sent within a distributed ledger rather than a centralized server-client system; hence the usage of blockchain in this situation specifically

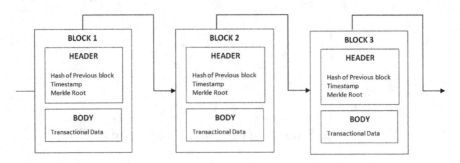

FIGURE 12.9 Basic structure of a blockchain.

resolves the criticality of data management and preserves it in a safe and immutable manner without tampering[58].

As a result, it is possible to continuously track changes in the condition brought on by outside sources. Smart contracts have made it possible for the two players to control orders automatically. The shop initiates a smart contract to send the order request. A password is given to both the sender and the receiver via the smart contract. This password will be connected to the producer's order, and the retailer will need to give it to the delivery driver in order to get the products. Thanks to the authentication on the blockchain platform with public and private key, the decision to accept or reject the order is necessary. As a result, the accesses are validated and certified, making it impossible to tamper with the system.

12.10 CONCLUSION

New business models will have a lot of chances as a result of the integration of the physical and digital worlds. A business process management strategy is required to assist practitioners and scholars in their comprehension of the possibilities of implementing advanced SC models because the introduction of smart technology is primarily driven by process modifications. This study fills the gap by synthesizing the usage of IoT applications in SCs in smart cities to increase visibility by conducting a literature study to evaluate existing research thoroughly. This work has looked in detail at how supply management processes are affected by the combined usage of technologies like RFID, IoT, and blockchain. It is implied from the considerable literature that the field of research fusing SCM with megatechnologies is increasing quickly. The study also explained that emergent technology can facilitate time savings in the present supply chain processes. The study also covered the advantages of sustainable SC traceability for the environment and the economy.

REFERENCES

1. D. M. Lambert, and M. C. Cooper, "Issues in supply chain management," *Ind. Mark. Manag.*, vol. 29, no. 1, pp. 65–83, 2000. https://doi.org/10.1016/S0019-8501(99)00113-3
2. R. Accorsi, S. Cholette, R. Manzini, and A. Tufano, "A hierarchical data architecture for sustainable food supply chain management and planning," *J. Clean. Prod.*, vol. 203, pp. 1039–1054, 2018. https://doi.org/10.1016/J.JCLEPRO.2018.08.275
3. A. Rejeb, S. Simske, K. Rejeb, H. Treiblmaier, and S. Zailani, "Internet of things research in supply chain management and logistics: A bibliometric analysis," *Internet Things*, vol. 12, p. 100318, 2020. https://doi.org/10.1016/J.IoT.2020.100318
4. M. Ben-Daya, E. Hassini, and Z. Bahroun, "Internet of things and supply chain management: A literature review", vol. 57, no. 15–16, pp. 4719–4742, 2017. https://doi.org/10.1080/00207543.2017.1402140
5. K. Butner, "The smarter supply chain of the future," *Strateg. Leadersh*, vol. 38, no. 1, pp. 22–31, 2010. https://doi.org/10.1108/10878571011009859
6. M. Christopher, and M. Holweg, "'Supply chain 2.0': Managing supply chains in the era of turbulence," *Int. J. Phys. Distrib. Logist. Manag.*, vol. 41, no. 1, pp. 63–82, 2011. https://doi.org/10.1108/09600031111101439
7. S. S. Kothari, S. V Jain, and P. Abhishek Venkteshwar, "The impact of IoT in supply chain management," *Int. Res. J. Eng. Technol.*, vol. 257, 2008. https://doi.org/10.1108/09574090310806503

8. Economic Intelligence Unit Report, "Companies without borders collaborating to compete," "An Economist Intelligence Unit report sponsored by BT", November 2006, Accessed: Oct 2023, Online, Available: http://graphics.eiu.com/files/ad_pdfs/eiu_BT_collaboration_wp.pdf.
9. B. L. Maccarthy, C. Blome, J. Olhager, J. S. Srai, and X. Zhao, "Supply chain evolution-theory, concepts and science," *Int. J. Oper. Prod. Manag.*, 2016. https://doi.org/10.1108/IJOPM-02-2016-0080
10. M. Casson, "Economic analysis of international supply chains: An internalization perspective," *J. Supply Chain Manag.*, vol. 49, no. 2, pp. 8–13, 2013. https://doi.org/10.1111/JSCM.12009
11. K. Woody, "Conflict minerals legislation: The SEC's new role as diplomatic and humanitarian watchdog," *Fordham Law Rev.*, vol. 81, no. 3, 2012.
12. G. Gereffi, "Global value chains in a post-Washington consensus world," *Rev. Int. Polit. Econ.*, vol. 21, no. 1, pp. 9–37, 2014. https://doi.org/10.1080/09692290.2012.756414
13. D. J. Ketchen, and L. C. Giunipero, "The intersection of strategic management and supply chain management," *Ind. Mark. Manag.*, vol. 33, no. 1, pp. 51–56, 2004. https://doi.org/10.1016/J.INDMARMAN.2003.08.010
14. F. Jia, R. Lamming, M. Sartor, G. Orzes, and G. Nassimbeni, "Global purchasing strategy and international purchasing offices: Evidence from case studies," *Int. J. Prod. Econ.*, vol. 154, pp. 284–298, 2014. https://doi.org/10.1016/J.IJPE.2013.09.007
15. S. Shah, K. Mokakangwe, B. Keitumetes, and S. Menon, "The growing adoption of internet of things on supply chain," *Int. J. Econ. Manag. Syst.*, vol. IV, pp. 108–112, 2019.
16. S. Ahmed et al., "Towards supply chain visibility using internet of things: A dyadic analysis review," *Sensors (Basel).*, vol. 21, no. 12, 2021. https://doi.org/10.3390/S21124158
17. D. Ivanov, and B. Sokolov, "Evolution of supply chain management (SCM)," *Adapt. Supply Chain Manag.*, pp. 1–17, 2010. https://doi.org/10.1007/978-1-84882-952-7_1
18. S. Cannella, R. Dominguez, J. M. Framinan, and B. Ponte, "Evolving trends in supply chain management: Complexity, new technologies, and innovative methodological approaches," *Complexity*, vol. 2018, 2018. https://doi.org/10.1155/2018/7916849
19. L. Wu, X. Yue, A. Jin, and D. C. Yen, "Smart supply chain management: A review and implications for future research," *Int. J. Logist. Manag.*, vol. 27, no. 2, pp. 395–417, 2016. https://doi.org/10.1108/IJLM-02-2014-0035
20. G. Zhang, Y. Yang, and G. Yang, "Smart supply chain management in Industry 4.0: The review, research agenda and strategies in North America," *Ann. Oper. Res.*, pp. 1–43, 2022. https://doi.org/10.1007/S10479-022-04689-1
21. M. C. Pedroso, and D. Nakano, "Knowledge and information flows in supply chains: A study on pharmaceutical companies," *Int. J. Prod. Econ.*, vol. 122, no. 1, pp. 376–384, 2009. https://doi.org/10.1016/J.IJPE.2009.06.012
22. L. Atzori, A. Iera, and G. Morabito, "The internet of things: A survey," *Comput. Netw.*, vol. 54, no. 15, pp. 2787–2805, 2010. https://doi.org/10.1016/J.COMNET.2010.05.010
23. M. Núñez-Merino, J. M. Maqueira-Marín, J. Moyano-Fuentes, and C. A. Castaño-Moraga, "Industry 4.0 and supply chain. A systematic science mapping analysis," *Technol. Forecast. Soc. Change*, vol. 181, p. 121788, 2022. https://doi.org/10.1016/J.TECHFORE.2022.121788
24. "Industry 4.0." https://www.gtai.de/en/invest/industries/industrial-production/industrie-4-0 (accessed Oct. 30, 2022).
25. S. A. Raza, "A systematic literature review of RFID in supply chain management," *J. Enterp. Inf. Manag.*, vol. 35, no. 2, pp. 617–649, 2022. https://doi.org/10.1108/JEIM-08-2020-0322
26. W. C. Tan, and M. S. Sidhu, "Review of RFID and IoT integration in supply chain management," *Oper. Res. Perspect.*, vol. 9, p. 100229, 2022. https://doi.org/10.1016/J.ORP.2022.100229

27. J. Yan et al., "Intelligent supply chain integration and management based on cloud of things," vol. 2014, 2014. https://doi.org/10.1155/2014/624839
28. S. Qrunfleh, and M. Tarafdar, "Supply chain information systems strategy: Impacts on supply chain performance and firm performance," *Int. J. Prod. Econ.*, vol. 147, no. PART B, pp. 340–350, 2014. https://doi.org/10.1016/J.IJPE.2012.09.018
29. C. N. Verdouw, A. J. M. Beulens, and J. G. A. J. van der Vorst, "Virtualisation of floricultural supply chains: A review from an internet of things perspective," *Comput. Electron. Agric.*, vol. 99, pp. 160–175, 2013. https://doi.org/10.1016/J.COMPAG.2013.09.006
30. A. Chandrashekar, and P. B. Schary, "Toward the virtual supply chain: The convergence of IT and organization," *Int. J. Logist. Manag.*, vol. 10, no. 2, pp. 27–40, 1999. https://doi.org/10.1108/09574099910805978
31. P. Bowman, J. Ng, M. Harrison, T. S. López, and I. Alexander, "Sensor based condition monitoring," "Building Radio frequency Identification for the Global Environment", European Union, June 2009, Accessed: Oct 2023, Online, Available: https://citeseerx.ist.psu.edu/document?repid=rep1&type=pdf&doi=620318c8da30de47afb70d0b69f1a273bdaed4c2.
32. L. He, M. Xue, and B. Gu, "Internet-of-things enabled supply chain planning and coordination with big data services: Certain theoretic implications," *J. Manag. Sci. Eng.*, vol. 5, no. 1, pp. 1–22, 2020. https://doi.org/10.1016/J.JMSE.2020.03.002
33. Z. Bi, L. Da Xu, and C. Wang, "Internet of things for enterprise systems of modern manufacturing," *IEEE Trans. Ind. Inform.*, vol. 10, no. 2, pp. 1537–1546, 2014. https://doi.org/10.1109/TII.2014.2300338
34. T. Wang, Y. F. Zhang, and D. X. Zang, "Real-time visibility and traceability framework for discrete manufacturing shopfloor," *Proc. 22nd Int. Conf. Ind. Eng. Eng. Manag. 2015*, pp. 763–772, 2016. https://doi.org/10.2991/978-94-6239-180-2_72
35. S. Alyahya, Q. Wang, and N. Bennett, "Application and integration of an RFID-enabled warehousing management system – A feasibility study," *J. Ind. Inf. Integr.*, vol. 4, pp. 15–25, 2016. https://doi.org/10.1016/J.JII.2016.08.001
36. M. Turrisi, M. Bruccoleri, and S. Cannella, "Impact of reverse logistics on supply chain performance," *Int. J. Phys. Distrib. Logist. Manag.*, vol. 43, no. 7, pp. 564–585, 2013. https://doi.org/10.1108/IJPDLM-04-2012-0132
37. D. Kiritsis, "Closed-loop PLM for intelligent products in the era of the internet of things," *CAD Comput. Aided Des.*, vol. 43, no. 5, pp. 479–501, 2011. https://doi.org/10.1016/J.CAD.2010.03.002
38. Y. Gu, and Q. Liu, "Research on the application of the internet of things in reverse logistics information management," *J. Ind. Eng. Manag.*, vol. 6, no. 4, pp. 963–973, 2013. https://doi.org/10.3926/jiem.793
39. G. C. Parry, S. A. Brax, R. S. Maull, and I. C. L. Ng, "Operationalising IoT for reverse supply: The development of use-visibility measures," *Supply Chain Manag.*, vol. 21, no. 2, pp. 228–244, 2016. https://doi.org/10.1108/SCM-10-2015-0386
40. H. Nissenbaum, "Privacy as contextual integrity," *Washingt. Law Rev.*, vol. 79, no. 1, pp. 119–158 Feb. 2004.
41. I. C. L. Ng, and S. Y. L. Wakenshaw, "The internet-of-things: Review and research directions," *Int. J. Res. Mark.*, vol. 34, no. 1, pp. 3–21, 2017. https://doi.org/10.1016/J.IJRESMAR.2016.11.003
42. J. Yu, N. Subramanian, K. Ning, and D. Edwards, "Product delivery service provider selection and customer satisfaction in the era of internet of things: A Chinese e-retailers' perspective," *Int. J. Prod. Econ.*, vol. 159, pp. 104–116, 2015. https://doi.org/10.1016/J.IJPE.2014.09.031
43. J. Zhang, S. K. Ong, and A. Y. C. Nee, "RFID-assisted assembly guidance system in an augmented reality environment," *Int. J. Prod. Res.*, vol. 49, no. 13, pp. 3919–3938, 2011. https://doi.org/10.1080/00207543.2010.492802

44. G. Putnik et al., "Smart objects embedded production and quality management functions," *Int. J. Qual. Res.*, vol. 9, no. 1, pp. 151–166, Mar. 2015.
45. K. L. Choy, G. T. S. Ho, and C. K. H. Lee, "A RFID-based storage assignment system for enhancing the efficiency of order picking," *J. Intell. Manuf.*, vol. 28, no. 1, pp. 111–129, 2014. https://doi.org/10.1007/S10845-014-0965-9
46. S. Mathaba, M. Adigun, J. Oladosu, and O. Oki, "On the use of the internet of things and Web 2.0 in inventory management," *J. Intell. Fuzzy Syst.*, vol. 32, no. 4, pp. 3091–3101, 2017. https://doi.org/10.3233/JIFS-169252
47. P. J. Reaidy, A. Gunasekaran, and A. Spalanzani, "Bottom-Up approach based on internet of things for order fulfillment in a collaborative warehousing environment," *Int. J. Prod. Econ.*, vol. 159, pp. 29–40, 2015. https://doi.org/10.1016/J.IJPE.2014.02.017
48. Z. Pang, Q. Chen, W. Han, and L. Zheng, "Value-centric design of the internet-of-things solution for food supply chain: Value creation, sensor portfolio and information fusion," *Inf. Syst. Front.*, vol. 17, no. 2, pp. 289–319, 2015. https://doi.org/10.1007/S10796-012-9374-9
49. "The Calculus of Fuzzy If/Then Rules | Proceedings of the Theorie und Praxis, Fuzzy Logik." https://dl.acm.org/doi/10.5555/646222.682494 (accessed Oct. 30, 2022).
50. N. Agami, M. Saleh, and H. El-Shishiny, "A fuzzy logic based trend impact analysis method," *Technol. Forecast. Soc. Change*, vol. 77, no. 7, pp. 1051–1060, 2010. https://doi.org/10.1016/J.TECHFORE.2010.04.009
51. E. Shekarian, E. U. Olugu, S. H. Abdul-Rashid, and E. Bottani, "A fuzzy reverse logistics inventory system integrating economic order/production quantity models," *Int. J. Fuzzy Syst.*, vol. 18, no. 6, pp. 1141–1161, 2016. https://doi.org/10.1007/S40815-015-0129-X
52. N. K. Sharma, V. Kumar, P. Verma, and S. Luthra, "Sustainable reverse logistics practices and performance evaluation with fuzzy TOPSIS: A study on Indian retailers," *Clean. Logist. Supply Chain*, vol. 1, p. 100007, 2021. https://doi.org/10.1016/J.CLSCN.2021.100007
53. H. Hasan, E. AlHadhrami, A. AlDhaheri, K. Salah, and R. Jayaraman, "Smart contract-based approach for efficient shipment management," *Comput. Ind. Eng.*, vol. 136, pp. 149–159, 2019. https://doi.org/10.1016/J.CIE.2019.07.022
54. P. Helo, and Y. Hao, "Blockchains in operations and supply chains: A model and reference implementation," *Comput. Ind. Eng.*, vol. 136, pp. 242–251, 2019. https://doi.org/10.1016/J.CIE.2019.07.023
55. V. G. Venkatesh, K. Kang, B. Wang, R. Y. Zhong, and A. Zhang, "System architecture for blockchain based transparency of supply chain social sustainability," *Robot. Comput. Integr. Manuf.*, vol. 63, p. 101896, 2020. https://doi.org/10.1016/J.RCIM.2019.101896
56. X. Yao, W. Du, X. Zhou, and J. Ma, "Security and privacy for data mining of RFID-enabled product supply chains," *Proc. 2016 SAI Comput. Conf. SAI 2016*, pp. 1037–1046, 2016. https://doi.org/10.1109/SAI.2016.7556106
57. M. A. Habib, M. B. Sardar, S. Jabbar, C. M. N. Faisal, N. Mahmood, and M. Ahmad, "Blockchain-based supply chain for the automation of transaction process: Case study based validation," *2020 Int. Conf. Eng. Emerg. Technol. ICEET 2020*, 2020. https://doi.org/10.1109/ICEET48479.2020.9048213
58. J. Rian Leevinson, V. Vijayaraghavan, and M. Dammodaran, "Blockchain mechanisms as security-enabler for industrial IoT applications," pp. 145–162, 2019. https://doi.org/10.1007/978-3-030-24892-5_7

13 Moving Toward Autonomous Vehicles (Drones and Robots) for Efficient and Smart Delivery of Services Using Hybrid Ontological-Based Approach

Nidhi, Jitender Kumar, and Sofia Sandhu

13.1 INTRODUCTION

A rapid growth and evolution is being seen in the logistics and transportation industries with respect to delivery services. With the advent of the e-commerce industry and modern technologies, there has been a rapid increase in the delivery of packages by various organizations. It is estimated that the current delivery of parcels is expected to grow around 262 billion packages by 2026 [1]. On the other hand, it has also raised customer expectations of getting their parcels and orders delivered at the doorstep at minimal cost. It urges the need to optimize and share transportation costs from one location to another across the city. Sharing of costs does not create problems until the packages are sorted and addressed to the receiver. But, as the package gets closer to the final destination, it doubles the initial cost incurred on transportation, which is commonly called as *last mile challenge*. The outcome is that till the package is received by a customer, it becomes the costliest and most complex delivery phase [2]. As a result, e-commerce businesses started altering the efficiency of their delivery policies to survive in the competitive market at reasonable prices. The solution to this complex delivery process is the employment of autonomous delivery technologies such as drones and robotics-based systems. These autonomous vehicles in this regard can provide a quick and flexible delivery service. Besides this, logistics using drones and robots can also reduce last mile shipping costs by reducing transportation costs. Some food joints in New Zealand have already started the use of robotics and drones in the delivery of food and medical services [3]. The

DOI: 10.1201/9781003353034-13

use of unmanned aerial vehicles (UAVs) also proved efficient during the COVID-19 pandemic in providing contactless delivery, thus mitigating the risk of transmission of virus among persons. Thus, the demand for contactless delivery modes and deployment of UAVs particularly for delivery of goods and medicines has risen exponentially during the health crisis. Also, the use of drones and robotics also contributes in reducing greenhouse emissions and global warming caused due to increasing transportation deliveries [4]. Besides this, drone-based delivery also reduces traffic congestion and the number of accidents on highly traffic routes [5]. Generally, the autonomous vehicles including drones and robots perform the delivery process in the manner as follows. Once the customer purchases goods from an online platform or some hyper store for doorstep delivery, a drone is assigned to pick up the ordered package. Based on the assignment process, an autonomous drone with operated batteries is being selected from the fleet on the basis of the target destination. After assignment, a robot is employed to pick up the package from the respective store and leaves the packet on drone area for initiation of the delivery process. Once the packet is delivered, the respective drone moves to the recharging station or waits for the next delivery request.

Since drones have limited flight range on the basis of battery capacity, it is mandatory to deploy recharging stations in order to increase the flight range and expedite the delivery process [6]. The chapter is organized into following sections. Section 13.2 provides comprehensive review of studies in respect of delivery services using drone and robotics-based logistics. Section 13.3 discusses methodology (individually as well as in hybrid mode) adopted for delivery process using autonomous ground vehicles (AGVs; robots) and autonomous aerial vehicles (AAVs; drones). Section 13.4 analyzes the performance of three systems in context of time intervals and waiting times with regard to live coordinates recorded during peak hours at Connaught Place (CP) in Delhi. Section 13.5 concludes the given chapter followed by future scope and references.

13.2 RELATED WORKS ON DRONE AND ROBOTICS-BASED LOGISTICS

The related works in the context of drone and robot logistics are broadly classified into four categories: vehicle routing problems in drones, assignment of specific drones for delivery, charging stations, and fleet dimensioning. Several studies are conducted in this context which is described as follows: Murray and Chu [7] dealt with issues related to the delivery of drones to minimize the optimization time of round trip. It handled the traveling salesman problem on the basis of mixed linear programming which can minimize the expected delivery time. This method is known as "Truck First, Drone Second," because the truck path is designed to solve a traveling salesman problem, while the drone is allowed to follow parallel scheduling. However, this approach does not consider the use of robots which can prove feasible in traffic-congested areas. Jeon et al. [8] utilize drones by reducing the number of flights using localization and data tracking. Dorling et al. [2] pointed out problems of delivery time and cost optimization by proposing an algorithm to optimize drone size and trips for delivery of packages. The authors assumed that the drones were fully chargeable and no deployment station was needed. It leads to the problem of swapping of batteries. Wang and coworkers [9, 10] proposed a "Drone vehicle traffic problem" with the aim of delivering packets in the shortest time. It used a combination of trucks and drones rather than robots and drones to ensure a cost reduction in delivery times by simulating through congested cities in worst-case scenarios. However, the authors did not propose any model or mechanism for delivery in hybrid mode. This problem is also covered in this chapter in further sections. Schermer et al. [11] in their study proposed heuristic algorithms in the context of solving vehicle routing problems and drone optimization. The study introduced the basic concept of robots but not as a tool for picking up delivery parcels. Another limitation in this study is the lack of a hybrid mode in the delivery process. Ham [12] in their study introduced delivery optimization scheduling to overcome delays in pick-up and delivery processes. Wang and Sheu [13] introduced the routing and optimization problem by using an arc-based approach and a price algorithm. This approach has the advantage that it can be extended to customers on a larger scale. Bertolaso et al. [14] introduced the concept of a hybrid delivery system which is the combination of drone and vehicle instead of robots. It does not mitigate the problem of traffic congestion on routes that is caused due to large number of vehicles. Grippa et al. [15] assigned requests to drones in order to reduce delivery time as much as possible. For the arrival of drone requests, this model uses a Poisson distribution and a shortest time-first scheduling algorithm. Yu et al. [16] addressed the issues related to drone recharging stations. The authors proposed an optimal path algorithm to visit multiple locations by drones without charging again and again. In their study, Shao et al. [17] proposed a swarm optimization and time scheduling approach for drone delivery services which includes charging batteries and maintenance checkpoints, in order to minimize the delivery time from pick-up location to destination. Alyassi et al. [18] employed the use of robots and drones for delivery services at an individual level and analyzed the delivery performance via empirical studies. The study also missed the hybrid last mile delivery approach. Baca et al. [19] made use of machine learning algorithms for the landing of drones on terraces of specific destinations. The authors did not give any information about robots as delivery agents. Feng et al. [20] proposed an autonomous landing method using filters and target localization for efficient delivery of packets. Hoffmann and Prause [21] discussed the legal issues related to the landing of drones but no real case study was incorporated to guarantee the effect of autonomous vehicles in parcel delivery services. A study conducted by Sunghun and Hyunsu [22] analyzed delivery by drones on the basis of cost and hardware used. Although, it did not introduce any hybrid mode of delivery to avoid traffic in highly congested areas. Goodchild and Toy [23] discussed the possibility of reducing greenhouse gas emissions by making use of UAVs. It is found that emissions vary across different delivery zones. Troudi et al. [24] investigated vehicle routing problem to address dimensioning problems of drone delivery services. The proposed approach used two strategies—one is scheduling delivery operations and the other is maintaining balance between drones and fleet size. An overview of the aforementioned literature studies is shown in Table 13.1.

TABLE 13.1
An Overview of Relevant Works on Drone and Robotics Logistics Systems

Source	Problem Addressed	Methods Used	Assumptions/Limitations
[7]	Drone delivery to minimize the optimization time of round trip	Solves traveling salesman problem by using Truck First, Drone second method	• Lack of real data for simulation • Does not deal with robots to avoid network congestion
[8]	Vehicle routing problem using drones.	Reduces number of flights using data localization	• No simulation parameters analyzed. • Drones cannot land on moving trucks
[2]	Delivery drones and cost optimization	Deals with cost issue and delivery time subject to cost limit	No provision for recharging batteries
[9, 10]	Drone vehicle traffic problem	Manages multiple drones at short intervals of time using heuristic optimization	No design introduced for hybrid mode
[11]	Vehicle routing problem and drone optimization	Used A* technique for calculation of traveling salesman problem using optimization algorithm	Introduced basic concept of robots but not for picking up delivery parcels
[12]	Vehicle routing problem with multiple drones and trucks	Solves scheduling problems of multiple trucks and drones	Drones perform both pick-up and delivery of parcels that causes less time to prepare food for nearby restaurants
[13]	Routing and optimization problem	Used an arc-based approach along with price algorithm to compute average waiting times of drones	• Does not consider flight range issue • Does not overcome the problem of road congestion
[14]	Vehicle routing and landing problem of drones	Used drones and vehicles instead of robots for delivery of packages	Employs limited number of customers
[15]	Drone assignment	Used Poisson distribution and shortest time first scheduling algorithm for assignment of drone requests	• No robots are assigned for pick-up process • Makes system complex by using Poisson distribution on random drone requests
[16]	Drone range and depot deployment issues	Drones are recharged from recharging stations with battery swapping feature	• No real data is used to validate the results • Unable to handle larger number of requests
[17]	Delivery times from pick-up to destination	Swarm optimization and time scheduling approach	Multiple swapping batteries lead to complex system design
[18]	Delivery criteria using robots and drones at an individual level	Empirical study and analysis	Missed the hybrid last mile delivery approach
[19]	Drone landing issues	Machine learning algorithms for landing drones on terraces	Lack of simulation and analysis to validate the proposed design

(Continued)

TABLE 13.1 (Continued)
An Overview of Relevant Works on Drone and Robotics Logistics Systems

Source	Problem Addressed	Methods Used	Assumptions/Limitations
[20]	Drone landing issues	Filter and localization of drones	Lack of simulation and analysis to validate the proposed design
[21]	Legal issues related to drones	Empirical and investigative method	No real case study to guarantee the effect of autonomous vehicles in delivery services
[22]	Delivery-related issues	Matrix function analysis	No real case study to guarantee the effect of autonomous vehicles in delivery services
[23]	Emission criteria across different delivery zones	Swarm optimization algorithm	No standards are inferred because of varying emission levels across different zones
[24]	Dimension problem of drones	Schedule delivery and maintenance approach	Consider drones for both pick-up and delivery services

13.3 METHODOLOGY

It is stated herewith that the delivery process in urban areas is one of the tedious tasks which is responsible for traffic and congestion on roads. The performance of transportation systems—one of the crucial factors for enhancing economic growth is deeply affected due to inefficient delivery processes. However, with new emerging technologies, viz. AGVs (robots) and AAVs (drones), the chapter proposes hybrid delivery processes, thereby mitigating the problem of congestion.

13.3.1 PROPOSED DELIVERY PROCESS USING AGVs (ROBOTS)

The given approach makes use of robots at an individual level. The robot system network comprises set of intersections and links. It is represented as R (I, L), where I denotes intersections and L denotes the links connecting various nodes.

Intersections consist of attributes namely, "x" and "y" where **x denotes pick-up location (herein referred to as food outlet)** and **y denotes destination (herein referred to as target location)** as set by the customer. The system also consists of **vehicles (V)** and **food parcels (FP)**, where vehicles signify **robots (R)**. The process is illustrated in Figure 13.1.

13.3.2 PROPOSED DELIVERY PROCESS USING ONTOLOGICAL-BASED HYBRID APPROACH (DRONES+ROBOTS)

The hybrid approach overcomes the problem of congestion and higher waiting time while delivering the packets at specific destinations. The hybrid approach is an amalgamation of two phases viz. robots used for picking up food packages and drones used for delivering food packages to the destination.

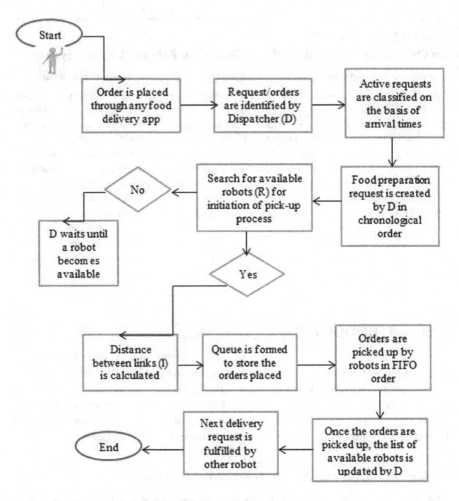

FIGURE 13.1 Food delivery process using robots

13.3.2.1 Phase 1

Robots (R) travels from **Depot (α)** to pick up food package from nearby **Restaurant (β)**. The robots are employed in order to avoid congestion on roads and they can travel sideways thus allowing the restaurant enough time to prepare the meal. It also makes use of one Robot Food Request Table (RFRT). In this phase, robots are only used to pick up from restaurant and drop it off back at the depot (α).

13.3.2.2 Phase 2

Drones are used to pick up packages from **depot (α)** and deliver to the **Destination (π)**. The reason behind this is the faster traveling speed of drones that overcomes the problem of congestion on road network thereby enhancing the delivery process at a faster rate and less waiting time. It makes use of **Drone Food Request Table (DFRT)**. The whole process is illustrated in Figure 13.2.

Moving Toward Autonomous Vehicles for Efficient and Smart Delivery

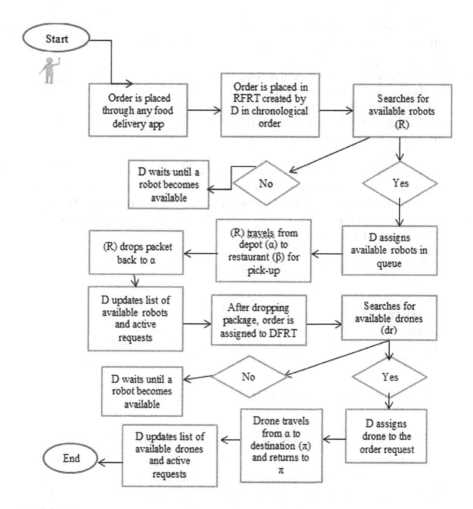

FIGURE 13.2 Food delivery process using hybrid approach

The algorithm for hybrid delivery process is as follows:

Algorithm: Pseudo code for food delivery process using hybrid approach

1. i=1;
2. Order is being placed by customer;
3. Dispatcher (D) is assigned;
4. "D" identifies the active requests in chronological order;
5. while active requests exists do
6. D creates RFRT on basis of order requests in chronological order;
7. D creates DFRT;
8. Searches for available robots (R) in the depot (α);

9. If R available at α
10. D assigns available robots in queue and travels from α to restaurant (β);
11. R drops packet back at α;
12. D updates the list of available robots at α;
13. D updates set of active requests in RFRT;
14. when order reaches α do;
15. D places order in queue in DFRT;
16. Search for available drones (dr) at α;
17. If dr available at α
18. D assigns available drones in queue and travels from α to destination (π);
19. A knowledge base is created for drones in the form of Ontology. Ontology is defined as a set of classes, properties and instances related to the given domain. It is used to increase re-usability and accessibility of the system;
20. D updates the list of drones;
21. else
22. D waits until drones become available at α;
23. end
24. else
25. D waits until a robot becomes available at α;
26. end
27. i=i+1;
28. end

The process is being referred in Figure 13.2.

13.3.3 Ontological Factor to Filter Delivery Orders in Hybrid System

After assignment of drones to fulfill the request of orders, drones are required to travel from depot to destination. After delivery, it may happen that one drone gets the order of two different destinations. In such cases, the location parameter needs to be filtered out. It arise the need to create some knowledge base or storage in the form of ontology that performs filtering of orders by location. It includes:

```
----Vehicle (Class)
-------Drones (Sub-class)
-----------D1, D2..........Dn (Instances)
----------------Location, OrderId (Parameters)
```

The knowledge base can either be created automatically or manual with the help of ontology editors like PROTÉGÉ, HOZO, etc.

13.4 CASE STUDY AND ANALYSIS

For analyzing the performance of three systems, the area of CP, Delhi is chosen due to its highly dense populated urban area and its latitude-longitudes are taken during the evening peak period from 6.00 pm to 7.00 pm [25].

The simulation model makes use of nodes, namely Restaurant (β), Depot (α), and Destination (π). Consider "X" website is used to identify nearby restaurants within 2 km of the boundary area from the source location. The location of depot (α) is chosen as Central Delhi because it signifies a hub or recharging station from where drones and robots depart for the delivery process. The simulation values are recorded using MATLAB and Simulink [26–28] and setup is done physically at the given location. Orders are created by matching nearby restaurant nodes and destination nodes through uniform distribution from 6.00 pm to 7.00 pm. A few scenarios have been investigated and noted with regard to delivery demand and the rise in frequency of orders.

In the robot scenario, accepting orders frequency is higher than getting it delivered because robots are allowed to wait until the previous order gets delivered and a new robot is assigned. The delivery performance is affected due to congestion on road networks. The delivery requests are not fulfilled as per the expectations of customers. Some orders are not delivered due to congestion and higher waiting times taken by robots and restaurants. The performance of delivery using robots is shown in Figure 13.3.

In the hybrid phase (robots and drones), it is seen that there are a higher number of orders delivered as compared to phase 1 and phase 2. The orders are picked up by the robots and left at a depot which in turn is picked up by the drones for delivery at the specific destination. The hybrid system can be said to

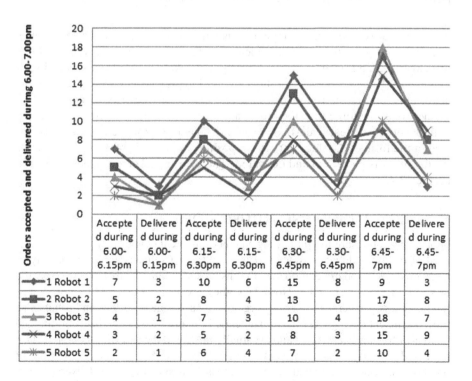

FIGURE 13.3 Delivery performance using robots

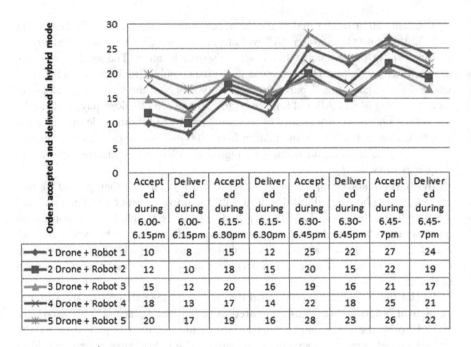

FIGURE 13.4 Delivery performance using drones and robots

be an alternative to individual logistics systems, thereby increasing the efficiency of delivery of packets at specified locations. This system mitigates the problem of congestion and ensures faster delivery of packets using drones traveling at higher speeds than robots. The robots are used to pick up and drop packets at the depot only so that the robots can pick up other order requests without wasting further time. By doing this, the orders are processed faster and drones can carry multiple orders to nearby destined locations. The drone knowledge base thus created in the form of ontology acts as a repository for storage of drones and robots till the order is picked up and delivered. The hybrid packet analysis is shown in Figure 13.4.

13.5 CONCLUSION AND FUTURE SCOPE

Drones and robots are the futuristic and emerging technologies that are defying the need of orthodox delivery systems. These automated vehicles enhance capacity of traditional fleets and are equipped with sensors and simulators that perform specific tasks in an efficient manner. Nowadays, they are in high demand by corporates, food business partners, and various organizations. Usage of drones and robots also proved efficient during COVID-19 pandemic in many ways, such that it minimized human-to-human interactions and eventually mitigated the risk of infections; it also increased capacity enhancement of delivery services when everything was out of order during pandemic. In addition to this, these automated vehicles delivered medical supply and goods even in highly congested areas.

In this chapter, three types of delivery systems are discussed and analyzed using the robot and drone technologies. The system includes AGV systems, AAV systems, and hybrid system that utilizes both robots and drones. AGV includes usage of a fleet of robots that run on pedestrians and sideways without interfering with ongoing road traffic network. AAV consists of fleet of drones operating and delivering food parcels at higher speeds. The third system is the hybrid system which is an amalgamation of ground vehicle system and aerial vehicle systems. In this system, robots are utilized for pick-up services in highly congested areas where drone landing is not feasible, and then package is placed in the drone, which is ready to be delivered from pick-up location to destination. A deep analysis of integration of autonomous and emerging technologies in the context of delivery systems is conducted by taking a place named CP into consideration. As it is a highly congested place, it will produce good simulation results when parameters and coordinates are taken during peak hours. Thus, whatever delay is incurred by robots in picking up goods is compensated by drone deployment as drone-related parcel delivery ensures sufficient networking resources for monitoring, safety, and operational purposes.

As a future research, the study can be extended to optimization and scheduling algorithms for assignment of order requests in terms of the shortest first arrival strategy. An optimization algorithm will help in finding optimal depot location which is nearby target locations. In turn it will enhance companies profit and reduces customer waiting time.

REFERENCES

1. J. J. Aurambout, K. Gkoumas, and B. Ciuffo, "Last mile delivery by drones: An estimation of viable market potential and access to citizens across European cities," European Transport Research Review, vol. 11, pp. 1–21, 2019.
2. K. Dorling, J. Heinrichs, G. M. Geoffrey, and S. Magierowski, "Vehicle routing Problems for Drone Delivery," **2016**, arXiv:1608.02305
3. M. S. Y. Hii, P. Courtney, and P. G. Royall, "An evaluation of the delivery of medicines using drones," Drones, vol. 3, p. 52, 2020.
4. A. Goodchild, and J. Toy, "Delivery by drone: An evaluation of unmanned aerial vehicle technology in reducing CO_2 emissions in the delivery service industry," Transportation Research, Part D: Transport and Environment, vol. 61, pp. 58–67, 2018.
5. M. Doole, J. Ellerbroek, and J. Hoekstra, "Estimation of traffic density from drone-based delivery in very low level urban airspace," Journal of Air Transport Management, vol. 88, pp. 101862–101862, 2020.
6. K. A. O. Suzuki, F. P. Kemper, and J. R. Morrison, "Automatic battery replacement system for UAVs: Analysis and design," Journal of Intelligent and Robotic Systems: Theory and Applications, vol. 65, pp. 563–586, 2021.
7. C. C. Murray, and A. G. Chu, "The flying sidekick traveling salesman problem: Optimization of drone-assisted parcel delivery," Transportation Research Part C: Emerging Technologies, vol. 54, pp. 86–109, 2015.
8. A. Jeon, J. Kang, B. Choi, N. Kim, J. Eun, and T. Cheong, "Unmanned aerial vehicle last-mile delivery considering backhauls," IEEE Access, vol. 9, pp. 85017–85033, 2021.
9. S. Poikonen, X. Wang, and B. Golden, "The vehicle routing problem with drones: Extended models and connections," Networks, vol. 70, pp. 34–43, 2017.
10. X. Wang, S. Poikonen, and B. Golden, "The vehicle routing problem with drones: Several worst-case results," Optimization Letters, vol. 11, pp. 679–697, 2017.

11. D. Schermer, M. Moeini, and O Wendt, "Algorithms for solving the vehicle routing problem with drones," In Intelligent Information and Database Systems, 2nd edition, Springer: Cham, Switzerland, 2018; Volume 10751, pp. 352–361.
12. A. M. Ham, "Integrated scheduling of m-truck, m-drone, and m-depot constrained by time-window, drop pickup, and m-visit using constraint programming," Transportation Research Part C: Emerging Technologies, vol. 91, pp. 1–14, 2018.
13. Z. Wang, and B. S. Sheu, "Vehicle routing problem with drones," Transportation Research Part B: Methodological, vol. 122, pp. 350–364, 2019.
14. A. Bertolaso, M. R. Masoume, A. Farinelli, and R. Muradore, "Using petri net plans for modeling UAV-UGV cooperative landing," Frontiers in Artificial Intelligence and Applications, vol. 285, pp. 1720–1721, 2016.
15. P. Grippa, D.A. Behrens, F. Wall, and C. Bettstetter, "Drone delivery systems: Job assignment and dimensioning," Autonomous Robots, vol. 43, pp. 261–274, 2018.
16. K. Yu, A.K. Budhiraja, and P. Tokekar, "Algorithms for Routing of Unmanned Aerial Vehicles with Mobile Recharging Stations," arXiv **2017**, arXiv:1704.00079v3. Available online: https://arxiv.org/abs/1704.00079
17. J. Shao, J. Cheng, B. Xia, K. Yang, and H. Wei, "A novel service system for long-distance drone delivery using the "Ant colony and A*" algorithm," IEEE Systems Journal, vol. 15, pp. 3348–3359, 2021.
18. R. Alyassi, M. Khonji, S. C. Chau, K. Elbassioni, T.C. Ming, and A. Karapetyan, "Autonomous Recharging and Flight Mission Planning for Battery-operated Autonomous Drones," arXiv 2017, arXiv:1703.10049.
19. T. Baca, P. Stepan, V. Spurny, D. Hert, R. Penicka, M. Saska, J. Thomas, G. Loianno, and V. Kumar, "Autonomous landing on a moving vehicle with an unmanned aerial vehicle," Journal of Field Robotics, vol. 36, pp. 874–891, 2019.
20. Y. Feng, C. Zhang, S. Baek, S. Rawashdeh, and A. Mohammadi, "Autonomous landing of a UAV on a moving platform using model predictive control," Drones, vol. 2, p. 34, 2018.
21. T. Hoffmann, and G. Prause, "On the regulatory framework for last Mile delivery robots," Machines, vol. 6, no. 3, p. 33, 2018.
22. J. Sunghun, and K. Hyunsu, "Analysis of amazon prime air UAV delivery service," Knowledge and Information Systems, vol. 12, no. 2, pp. 253–266, 2017.
23. A. Goodchild, and J. Toy, "Delivery by drone: An evaluation of unmanned aerial vehicle technology in reducing CO_2 emissions in the delivery service industry," Transportation Research, Part D: Transport and Environment, vol. 61, pp. 58–67, 2018.
24. A. Troudi, S. A. Addouche, S. Dellagi, and A. El Mhamedi, "Sizing of the drone delivery fleet considering energy autonomy," Sustainability, vol. 10, p. 3344, 2018.
25. https://www.findlatitudeandlongitude.com/l/Connaught+Place%2C+New+Delhi-110001/394735/
26. https://in.mathworks.com/discovery/drone-simulation.html
27. V. A. Budnyaev, I. F. Filippov, V. V. Vertegel, and S. Y. Dudnikov, "Simulink-based Quadcopter Control System Model," 2020 1st International Conference Problems of Informatics, Electronics, and Radio Engineering (PIERE), 2020, pp. 246–250. https://doi.org/10.1109/PIERE51041.2020.9314676
28. P. Mittal, and R. Singh, "A simulated dataset in aerial images using Simulink for object detection and recognition," International Journal of Cognitive Computing in Engineering, vol. 3, 2022, 144–151, ISSN 2666-3074. https://doi.org/10.1016/j.ijcce.2022.07.001

14 Relevance and Predictability in Wireless Multimedia Sensor Network in Smart Cities

Raj Gaurang Tiwari, Pratibha, Sandip Vijay,
Sandeep Dubey, Ambuj Kumar Agarwal,
and Megha Sharma

14.1 INTRODUCTION

Nowadays, multimedia application over wireless sensor network (WSN) is in great demand and it becomes a necessary part of everyone's day-to-day life in the era of smart cities. Due to certain limitations of WSN, such as battery issues, security issues, low communication speeds, etc., multimedia sensors are getting more attention from researchers. The most beneficial thing is that multimedia sensors are capable of exquisite video, audio, image, and scalar sensor data and transfer multimedia data via the sensor network.

WSNs consist of up to thousands of sensor nodes that work together to detect varying conditions like sound, motion, temperature, vibrations, pressure, etc. at distinct sites using IoT sensors [1, 2]. There are many applications where WSNs work well like target tracking, medical applications, security surveillance, habitat detecting, etc. There is more probability of data loss over WSNs at the time of data transfer than over wired networks. In wired networks, data get lost generally due to congestion problems; however, in WSN, there are many numbers of reasons due to why data get lost like node failure, habitat interface, noise, unreliable link, unexpected damage, etc. [3].

Due to the limitations of inherent characteristics of sensor nodes communication over WSN is not so reliable. Whereas reliability is the main characteristic of a quality-orientated WSN and it becomes more difficult to attain reliable communication due to the ubiquity of wireless links with greater inaccuracy rate. There are some more challenges introduced due to the demand of multimedia application requirements like time constraints, high bandwidth, and varying reliability requirements [4, 5]. These are key factors to outline communication protocol to get a proficient multimedia transmission in sensor networks.

Figure 14.1 shows a traditional WSN. A sink node or base station confederates in between the network and users. Normally a WSN network carries thousands of sensor nodes and all sensor nodes can easily interact with each other via radio signals.

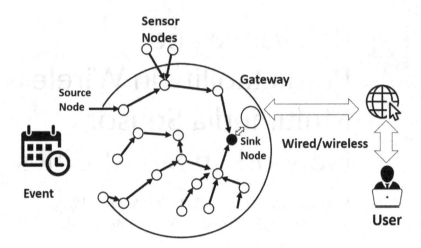

FIGURE 14.1 Conventional wireless sensors network.

These nodes are well attired with advanced computing and sensing devices, power components, and radio transceivers. A sensor node transmits the message including the image to the sink node when it finds out about a highly prior event [6, 7]. Urgent messages have greater significance with the least end-to-end deferral, whereas fewer prior messages have a large end-to-end delay. Data that get delayed at the receiver will be discarded, and it cannot be displayed. So, there is a need to use an advanced transport protocol that is suitable and convenient for multimedia communication and can handle all restrictions of sensor nodes.

The main objective to use an advanced transport protocol is to get better reliable and energy-efficient real-time multimedia data transmission over WSN and to achieve the extended life period of the network. To avoid congestion, a congestion control mechanism is also required to get better quality multimedia data [8].

The subsequent part of this chapter is arranged as follows: Section 14.2 throws light on diverse work done in the field of transport protocols. In Section 14.3, the transport protocols are discussed. Section 14.4 compares existing protocols. Section 14.5 concludes the chapter. In Section 14.6, possible future works are suggested.

14.2 RELATED WORKS

Real-time multimedia transmission over wired/wireless networks/WSN is already possible with a number of existing protocols. It was proposed that the real-time transport protocol (RTP) [9] be used for the transmission of multimedia data across an IP-based network. It does not allow for the fragmentation or re-assembly of traffic. Implementing RTP over a WSN is not a viable solution due to the high electricity consumption of RTP.

The Multi-flow Realtime Transport Protocol (MRTP) is an improvement on the present RTP [10]. RTP is used in conjunction with the multipath transmission. MRTP is a session-oriented protocol that can be used to determine data flow.

For WSN, numerous protocols for transmitting multimedia data have been developed. For effective multimedia transmission in WSN, Puga. [11] suggests Multimedia Distributed Transport for Sensor Networks (M-DTSN). This network assesses the transmission of a specific amount of data and determines whether the channel environment is enough to complete the transmission. M-DTSN evaluates a variety of characteristics in order to achieve effective transmission. Two of the factors that have been explored include time-consuming multimedia flows and a sensor that is not very energy efficient for transmission. As a result, there is no jamming control mechanism, which is necessary to get rid of any jamming.

Zakaria and El-Marakby [12] proposed a protocol named AdamRTP which is comparable to MRTP. MRTP is a session-oriented protocol designed for ad hoc networks, whereas AdamRTP is designed for WSN. With the help of the MDC encoder, AdamRTP separates the multimedia source stream into minor autonomous flows and transmits data for each flow to the receiving point through a joint/disjoint channel.

In WSN, Qaisar and Radha [13] proposed a protocol named MMDR, which employs three techniques to improve the received video quality. Source-coded video is divided into prioritized streams, with video packets routed via path diversity and data packets partially decoded.

Many protocols, such as MMDR, include dependability mechanisms but do not take into account jamming control mechanisms. The jamming control system is required to reduce energy ingestion during multimedia data transfer. In the future, energy efficiency should be the primary criterion for WSN transport protocols, particularly for multimedia data, which requires a large bandwidth and real-time latency due to the sensor nodes' short lifespan. A transport protocol is required that can consistently convey messages, control congestion in the most energy-efficient manner, and handle multimedia needs.

Sensor transmission control protocol (STCP), event to sink transport protocol (ESRT), and asymmetric and reliable transport protocol (ARTP) are some of the protocols that focus on both reliability and congestion control (ART). They do, however, have some flaws and room for growth [14].

The majority of the functions in the STCP [15] are implemented at the base station. STCP allows multiple programs to share a single network and delivers reliable congestion detection. The management of congestion information relies heavily on data packets. SenTCP has several performance issues, but STCP solves them by assuming that all sensor nodes in a WSN have synchronized clocks. STCP is a transport layer protocol that is widely used, scalable, and dependable.

The ESRT [16] protocol is one of the novel transport options. ESRT can accurately identify events in a WSN while consuming the least amount of energy. The ESRT's purpose is to manage the reporting occurrence of source nodes in order to deliver the desired end-to-end consistency guarantee while consuming the least amount of energy.

The main characteristics of ART [3] are distributed congestion control, reliable query transfer, and unswerving event transfer. Essential nodes (E-nodes) are a subset of sensor nodes that cover the full area to be sensed in an energy-efficient manner, whereas nonessential nodes (N-nodes) cover the complete area to be felt. If there is no jamming, the E-node and N-node are both communicating with the sink. The downside of ART is that if any packet is lost due to congestion at any N-node and is not recognized, there is no guarantee of recovery because E-node manages two-way reliability and congestion control.

Apart from the aforementioned convergence of the latest technologies like blockchain, machine learning, and augmented and virtual reality in a smart city over 4G/5G networks, this is not possible with fast data and continuous data communication. Several latest contributions have designed and implemented smart cities services over wireless multimedia sensor networks [17, 18].

14.3 TRANSPORT PROTOCOL

Transport protocol's work is to make the connection between the application layer and network layer, so that data can be delivered between source and target reliably it also manages congestion control and synchronizes heavy traffic suddenly inserted into the network. Generally, a huge amount of real-time data are incorporated with multimedia systems and sources. This leads to the need for a high bit rate to maintain communication quality. So, a protocol is required which meets both the demands of communication through multimedia and is compatible with sensor node characteristics [19–21].

14.3.1 RELIABILITY MECHANISM

Depending on the WSN application, different levels of reliability may be required. Communication takes place between sensor nodes and sink nodes in a WSN, and the direction of communication should be distinguished. In upstream communication, sensor nodes are senders, sink nodes (sensor-to-sink) are receivers, and vice versa in downstream communication [22].

14.3.2 PERFORMANCE METRICS

To evaluate the performance of any transport layer protocol some metrics like reliability, congestion, and energy efficiency can be used.

14.3.2.1 Reliability

Upstream (sensor-to-sink) and downstream (sink-to-sensor) reliable data transmission can meet reliability metrics (sink-to-sensor). WSNs have two forms of dependability: event and packet.

Package dependability is defined as the successful delivery of all packets or a success ratio [23]. Event dependability is defined as the success of an event report [24].

$$\sum_{t=1}^{T} \frac{\text{Probability of success of } m_t}{T}$$

where T is the total events, m is a message, t is the event, and it is the message that contains t.

$$\text{Node reliability of node } j, \left[\text{Rn}(j)\right] = \frac{\text{No. of packets received by the sink at } j}{\text{Total no. of packets generated at } j}$$

14.3.2.2 Congestion Degree

$$d(j) = \frac{t_s^j}{t_a^j}$$

where d is the jamming detection metric [24], t_s^j is the time taken of mean packet servicing, and t_a^j is the time taken of mean packet inter-arrival of node j.

Node efficiency or average delivery ratio =

$$\frac{\text{No. of packets received by } j}{\text{No. of packets received by } j\text{'s previous node}}$$

14.3.2.3 Analysis of Real-Time Multimedia Transmission

As the energy of sensor nodes is very restricted so it is necessary for the transport protocol to preserve extraordinary energy to make the most of the lifetime of the system. In WSN, packet loss can be frequent because of bit error or jamming. Thus,

$$\text{Packet loss ratio} = \frac{\text{No. of lost packets}}{\text{No. of packets produced by sensing nodes}}$$

Let us assume that dropped packets are directly related with energy depletion [24],

The energy loss per node $[E(j)] = \dfrac{\text{Total number of packets dropped by node } j}{\text{Total number of packets received by node } j}$

The energy loss for the whole network $[E_{network}] =$

$$\frac{\text{Total number of packets dropped by network}}{\text{Total number of packets received by the sink}}$$

The remaining energy in a sensor node, $E_r = \dfrac{\text{Remaining energy}}{\text{Initial energy}}$

The packet delivery rate is essential to assess data transmission performance since packet loss can take place at any step of the network owing to jamming.

$$\text{Packet delivery rate} = \frac{\text{No. of packets received by sink}}{\text{No. of packets generated by the source}}$$

"Peak signal-to-noise ratio (PSNR)" is utilized to determine the worth of video sequences at the decoder [25]:

$$\text{PSNR} = 20\log\frac{255}{\sqrt{MSE}}$$

where MSE is a mean square error.

14.3.2.4 Data Transmission in accordance with Data Type

Generally, a huge amount of real-time data are incorporated with multimedia systems and sources [26]. Due to this requirement, a high bit rate is generated to maintain quality and application. So the protocol is required which meets both demands of communication through multimedia and is compatible with sensor node characteristics [27–29].

Images ("I, P, and B frames"), video, and audio are all divided into frames in a multimedia data stream like MPEG server or video encoder. Each frame has a different level of trustworthiness. Following this level, data communication is limited to image transmission. The PR-SCTP protocol can be used to implement a new suggested transport protocol that provides various levels of dependability [30]. Data are transmitted in three streams since there are three varieties of image data (see Figure 14.2). I-frame transmission is signified by Stream 1, which has the highest priority. P-frames and B-frames up to the next I-frame are worthless if I-frames are damaged or lost during data transfer [31]. This transmission can be split into two parts: queue priority and data transmission differentiation. Here, the concentration is on data transmission based on data category.

The multimedia data transmission flowchart in Figure 14.3 is based on the data type. PR-SCTP and TC-SCTP are used for these protocols. In order to avoid unwanted data transmission, PR-SCTP changes transmission according to the data priorities by decreasing the repetition rate of every data. Before sending the data, TC-SCTP does the same purposeful check on each data's transmission time. The aforementioned protocols for the wired network, therefore, require a new protocol to use the IEEE 802.15.4 WSN algorithm.

This protocol is used to limit TC-SCTP duration and decide whether data on the recipient's side can be played on the buffer [32]. Only a single buffer is used in this suggested protocol and unique time limits are not required. The transmitting time is assigned with the assessment of the maximum time allowed for all data played at the recipient part of the transmitter.

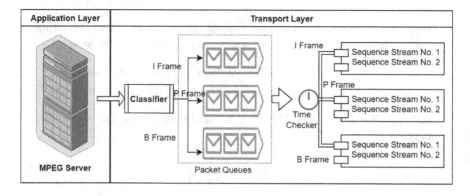

FIGURE 14.2 Multimedia data transmission in accordance with the type of data.

Relevance and Predictability in Wireless Multimedia Sensor Network

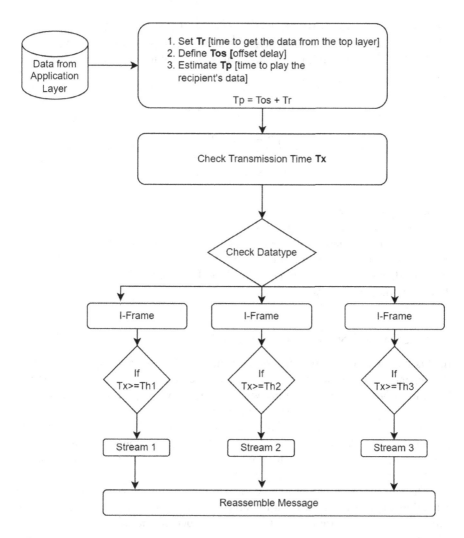

FIGURE 14.3 Flowchart of multimedia data transmission.

Tp is the time to play the recipient's data, Tos is the offset delay and Tr is the time to get the data from the top layer. The Tos is defined as the greatest amount of time to receive data on the recipient side from the top layer. Therefore,

$$Tp = Tos + Tr$$

To certify unswerving data transmission between source and target, unnecessary transmission is avoided, so that the buffer controls transmission time, T_x for transferring all data before transmission. The data are successfully transferred to the recipient if the transmission time is long. If not, the transmitter will not continue transmitting and deleting buffer data.

The transmission time, $T_x = Tp - T$

When transmission time is calculated, the type of data is checked whether it is I-frame, P-frame, or B-frame to reach the condition of transmission time T_X.

Th1, Th2, and Th3 are the threshold values decided using retransmission time-out (RT).

$$Th1 = (RT \times 2) + a_1$$

$$Th2 = (RT \times 1) + a_2$$

$$Th3 = (RT \times 2) + a_3$$

where a_1, a_2, and a_3 are the constants.

If data are I-frame, then T_x is greater than the Th1, after that data will be transferred to stream1. If data are P-frame, then T_x is greater than the Th2, after that data will be transferred to stream2. If data are B-frame, then T_x is greater than the Th3, after that data will be transferred to stream3. If T_x is smaller than Th3, then data are dropped without transmission.

14.4 COMPARISON OF PROTOCOLS

In this part, certain present transport protocols have been compared based on the dependability of message delivery, energy efficiency, and jamming control. Existing protocols cannot comply with all needs. Many protocols do not focus on reliability methods and just on congestion control. But three of the current protocols offer both transport layer, reliability, and capabilities for congestion control [33–36]. Almost all protocols use the implied congestion notification methodology, the queue length procedure for the congestion detection stage, and the exact rate adjustment method. Children's nodes can lower transmission rates to prevent more parent node congestion by changing rates [11]. The exact rate adjustment is a well-known technique in which the node schedules packet transmission at specified times to meet the expected rate to make special rate adjustments [11].

The most critical function at the transport layer is dependability, which ensures that data are delivered correctly from the source to the destination or sink node. Because the bulk of proposed protocols is application-based, there are numerous dependability strategies accessible. Aside from that, many protocols employ negative acknowledgment (NACK) and timeout to detect loss, as well as notification and packet retransmission to recover from the loss. In the future, protocols that are interoperable with a wide range of applications should be developed. The present transport protocol, which provides a dependability mechanism, is summarized in Table 14.1. The summary of the present transport protocol that provides a jamming control is revealed in Table 14.2.

Because sensor nodes have such a short operating system life, most protocols do not prioritize energy conservation. As a result, achieving energy efficiency in WSN is critical.

TABLE 14.1
Existing Transport Protocols Reliability Mechanism and Energy Efficiency Comparison

Protocol Name	Category	Type	Direction	Detection and Notification of Loss			Recovery of Loss		
				ACK	NACK	Time out	Increase Source Sensing Rate	Packet Retransmission	Energy Efficiency
SCTP	Frames	End to end	Both	Yes	Yes	Yes		Yes	No
ERTP	Packet	End to end	Upstream			Yes		Yes	Yes
GARUDA	Packet	Hop by hop	Downstream	Yes				Yes	No
PSFQ	Packet	Hop by hop	Downstream	Yes	Yes			Yes	No
STCP	Event packet	End to end	Upstream	Yes	Yes	Yes		Yes	No
ESRT	Event	Event to sink	Upstream			Yes	Yes		Yes
ART	Event query	End to end	Both	Yes	Yes	Yes		Yes	No

14.5 CONCLUSION

In this chapter, various transport protocols have been investigated. Also, existing transport protocols have been compared for real-time applications. The suggested transport protocol aimed to provide a dependable data transfer system as well as an effective congestion control system. Congestion control and dependability issues

TABLE 14.2
Existing Transport Protocols Congestion Control Comparison

Protocol Name	Packet Sending Success	Service Time	Queue Length	Notification of Congestion		Rate Adjustment	
				Explicit	Implicit	Exact Rate Adjustment	Additive Increase Multiplicative Decrease (AIMD)
SCTP	End to end	Yes	Yes		Yes		
ERTP							
GARUDA							
PSFQ							
STCP	End to end		Yes		Yes		Yes
ESRT	End to end		Yes		Yes	Yes	
ART	End to end	Yes			Yes		

will undoubtedly aid in lowering packet loss, resulting in a more energy-efficient overall system over WSN. The proposed protocol was able to meet the requirements for both multimedia communications over WSN and WSN characteristics. The primary issues that are required to be deliberated for the upcoming transport protocol in WSNs are high bandwidth demand, energy efficiency, minimal real-time delay, processing, and limited memory.

14.6 FUTURE SCOPE

The suggested transport protocol can be used on network simulator 2 to calculate the protocol's performance in the future. MPEG-4 video encoder can be exploited to send data via IEEE 802.15.4 network. The cross-layer optimization algorithm could be an additional aspect in improving the proposed transport protocol's efficiency. Cross-layer optimization also aids in network trustworthiness and eminence improvement. Different layers of the network can collaborate to share information, which helps to keep energy usage low.

REFERENCES

1. D. Kandris, C. Nakas, D. Vomvas and G. Koulouras. 2020. Applications of wireless sensor networks: An up-to-date survey. Applied System Innovation, 3(1): 14.
2. H. Agarwal, P. Tiwari and R. G. Tiwari. 2019, December. Exploiting Sensor Fusion for Mobile Robot Localization. In 2019 Third International conference on I-SMAC (IoT in Social, Mobile, Analytics and Cloud) (I-SMAC) (pp. 463–466). IEEE.
3. R. Rana and R. Kumar. 2019. Performance analysis of AODV in presence of malicious node. Acta Electron Malaysia, 3(1): 01–05.
4. A. K. Agarwal, R. G. Tiwari, V. Khullar and S. Dutta. 2021, September. Swarm Inspired Artificial Bee Colony Algorithm for Clustered Wireless Sensor Network. In 2021 9th International Conference on Reliability, Infocom Technologies and Optimization (Trends and Future Directions) (ICRITO) (pp. 1–5). IEEE.
5. R. G. Tiwari, M. Husain, V. Srivastava and K. Singh. 2011. "A Hypercube Novelty Model for Comparing E-Commerce and M-Commerce." In Proceedings of the 2011 International Conference on Communication, Computing & Security, pp. 616–619.
6. L. L. Hung, F. Y. Leu, K. L. Tsai and C. Y. Ko. 2020. Energy-efficient cooperative routing scheme for heterogeneous wireless sensor networks. IEEE Access, 8: 56321–56332.
7. A. Kumar, R. G. Tiwari, N. Kumar Trivedi, A. Anand, A. Kumar Agarwal and D. Prasad. 2021. "Extended Network Lifespan with Fault-Tolerant Information Transmission." In 2021 10th International Conference on System Modeling & Advancement in Research Trends (SMART), pp. 218–222. IEEE.
8. A. A. Ahmed. 2008. Secure Real-time Routing Protocol for Wireless Sensor Network. Phd Thesis, Universiti Teknologi Malaysia.
9. H. Schulzrinne, S. Casner, R. Frederick and V. Jacobson. 2003. RTP: A Transport Protocol for Real-Time Application. RFC 3550.
10. S. Mao, D. Bushmitch, S. Narayanan and S. S. Panwar. 2006. MRTP: A multiflow real-time transport protocol for ad hoc network. IEEE Transaction on Multimedia, 8(2), pp.356–369.
11. J. F. M. Puga, G. M. Fernandez, A. Griloand and N. M. C. Tiglao. 2010. Efficient Multimedia Transmission in Wireless Sensor Network. In Proceeding 6th Euro NF Conference on Next Generation Internet, NGI.

12. A. Zakaria and R. El-Marakby. 2009. AdamRTP: Adaptive Multi-flows Real-time Multimedia Delivery over WSNs. In IEEE International Symposium on Signal Processing and Information Technology, ISSPIT 2009.
13. S. B. Qaisar and H. Radha. 2009. Multipath Multi-Stream Distributed Reliable Video Delivery in Wireless Sensor Network. In Proceedings - 43rd Annual Conference on Information Sciences and Systems, CISS 2009.
14. D. Meneses, A. Grilo and P. R. Pereira. 2011. A Transport Protocol for Real-time Streaming in Wireless Multimedia Sensor Network. In Proceeding 7th EURO-NGI Conference on Next Generation Internet Networks, NGI.
15. Y. Iyer, S. Gandham and S. Venkatesan 2005. STCP: A Generic Transport Layer Protocol for Wireless Sensor Networks. In Proceedings of IEEE ICCCN 2005. San Diego, CA, USA, October.
16. O. B. Akan and I. F. Akyildiz. 2005. Event-to-sink reliable transport in wireless sensor network. IEEE/ACM Transaction on Networking, 13: 5.
17. R. Kumar, R. C. Singh and R. Khokher. 2022. Framework for Modeling, Procuring, and Building Systems for Smart City Scenarios Using Blockchain Technology and IoT, The Data-Driven Blockchain Ecosystem, CRC Press, pp. 30–50 https://doi.org/10.1201/9781003269281-3
18. E. U. Ogbodo, A. M. Abu-Mahfouz and A. M. Kurien. 2022. A survey on 5G and LPWAN-IoT for improved smart cities and remote area applications: From the aspect of architecture and security. Sensors, 22(16): 6313. https://doi.org/10.3390/s22166313
19. F. Yunus, N. S. N. Ismail, S. H. S. Ariffin, A. A. Shahidan, N. Fisal and S. K. S. Yusof. 2011. Proposed Transport Protocol for Reliable Data Transfer in Wireless Sensor Network (WSN). In Fourth International Conference on Modelling, Simulation and Applied Optimization.
20. C. Wan, A. T. Campbell and L. Krishnamurthy. 2002. PSFQ: A Reliable Transport Protocol for Wireless Sensor Network. 1st ACM International Workshop on Wireless Sensor Network and Application.
21. S. J. Park, R. Vedantham, R. Sivakumar and I. F. Akyildiz. 2004. A Scalable Approach for Reliable Downstream Data Delivery in Wireless Sensor Network. In Proceeding of ACM MobilHoc'04.
22. B. Yang, Reliable Data Delivery in Wireless Sensor Network, Master Thesis, University of Saskatchewan, Canada, 2010.
23. C. Wang, K. Sohraby, V. Lawrence, B. Li and Y. Hu. 2006. Priority-based Congestion Control in Wireless Sensor Networks. In Proceeding of IEEE International Conference on Sensor Networks. Ubiquitous and Trustworthy Computing (SUTC'06).
24. M. A. Rahman, A. E. Saddik and W. Gueaieb. 2008. Wireless Sensor Network Transport Layer: State of the Art. Sensors, Berlin Heidelberg: Springer-Verlang.
25. A. Argyriou. 2005. A Novel End-to-End Architecture for H.264 Video Streaming Over the Internet. Telecommunication System, Springer Science + Business Media.
26. Z. Lifen, S. Yanlei and L. Ju. 2003. The performance study of transmitting MPEG4 over SCTP. Proceeding of Neural Network and Signal Processing, 2: 1639–1642.
27. F. Stann and J. Heidemann. 2003. RMST: Reliable Data Transport in Sensor Networks. IEEE International Workshop on Sensor Net Protocol and Applications (SNPA).
28. H. Zhang, A. Arora, Y. Choi and M. G. Gouda. 2005. Reliable BurstyConvergecast in Wireless Sensor Networks. In Proceeding ACM MOBIHOC'05.
29. T. Le, W. Hu, P. Corke and S. Jha. 2009. ERTP: Energy-efficient and reliable transport protocol for data streaming in wireless sensor networks. Journal of Computer Communication, 32: 1154–1171.
30. R. Stewart, K. Morneault, C. Sharp, H. Schwarzbauer, T. Taylor, I. Rytina, M. Kalla, L. Zhang and V. Paxson. 2000. Stream Control Transmission Protocol. IETF RFC 2960.

31. N. S. N. Ismail, F. Yunus, S. H. S. Ariffin, A. A. Shahidan, R. A. Rashid, W.M.A.E.W. Embong, N. Fisal and S. K. S. Yusof. 2011. MPEG-4 Video Transmission using Distributed TDMA MAC Protocol over IEEE 802.15.4 Wireless Technology. In Fourth International Conference on Modelling, Simulation and Applied Optimization.
32. K. H. Kim, K. M. Jeong, C. H. Kang and S. J. Seok. 2009. A transmission control SCTP for real-time multimedia streaming. Journal of Computer Network, 54: 1418–1425.
33. C. Y. Wan, S. B. Eisenman and A. T. Campbell. 2003. CODA: Congestion Detection and Avoidance in Sensor Networks. In Proceeding of ACM Sensys'03.
34. C. Wang, K. Sohraby and B. Li. 2005. SenTCP: A hopby-hop Congestion Control Protocol for Wireless Sensor Networks. In Proceeding of IEEE INFOCOM.
35. B. Hull, K. Jamieson and H. Balakrishnan. 2004. Mitigating Congestion in Wireless Sensor Networks. In Proceeding ACM Sensys'04.
36. C. T. Ee and R. Bajcsy. 2004. Congestion Control and Fairness for Many-to-One Routing in Sensor Networks. In Proceeding ACM Sensys'04.

Index

A

Accuracy, 164
AdamRTP, 254
Agent, 94, 96
Algorithm(s), 13, 16, 163, 166–168, 170
Amazon simple storage, 52
AMQP, 47, 49
Analytics, 184
Anonymity, 163, 166–168, 170–171
API, 205
Arduino UNO 203, 204, 207, 208, 209, 212
ARM CPU, 206
Artificial Intelligence (AI), 2, 16–17, 24, 62, 83, 89, 100, 117
ARTP, 254
ASEAN Smart Cities Network (ASCN) 88
Association of Southeast Asian, 88; see also ASEAN
ATM, 94, 96
Attacks, 162, 166, 168, 170
Authentication, 167–169
Authenticity, 168, 170
Authority, 163, 165–166, 171
Automatic room light controller, 205
Automation, 26, 28, 32, 34, 37
Autonomous aerial vehicles, 240
Autonomous driving, 117; see also Self-driving car
Autonomous ground vehicles, 240
Autonomous vehicles, 239
AWS, 205, 216

B

Back propagation, 130
Big data, 17, 83–85, 87–89, 91, 93, 94–101
Big data analytics, 204
Bigdata, 196
Bigdata analytics, 180, 196
Bitcoin (BTC), 83, 163, 166
Blackhole attack, 111
Block, 162–165, 167–169, 171
Blockchain, 28, 30, 32, 34–37, 45, 162–171, 221, 234, 235
Blockchain system, 84, 85
Blockchain technology, 83–87, 89–97, 99–101
Bluetooth 45, 205, 214

C

CAN bus, 206
Car detection, 14
Car innovations, 12
CCTV, 17
CDMA, 45
Cellular area network, 45
Change in global population in rural and urban from 1950 to 2050, 73
Cloud, 30–32, 46, 52, 126
Cloud computing, 17, 85, 91, 137–138, 141, 194, 204
Cloud infrastructure, 16
Cloud storage, 194
Collision-avoidance systems, 23
Communications, 26, 27, 37
Completely invasive attacks, 142, 144
Computer vision, 12
Concept of smart city model, 59
Conclusion, 76
Confidentiality, 168, 171
Congestion degree, 256
Consensus, 163, 166–167
Constant Acceleration Heuristics (CAH), 9–10
Control, 205, 206, 207, 209, 214, 215
Controller board, 207
Convergence, 164, 166, 168, 170
Convolutional neural network (CNN), 129, 130
Cooperative Adaptive Cruise Control (CACC), 11
Correlations, 186
Cryptocurrency/Cryptocurrencies, 83, 85, 100, 163–164
Cryptographic hashing, 149–150
Cryptography, 162, 167–168, 170–171
Cyber-physical, 84, 86, 87, 100; see also Cyber-physical system (CPS)
Cyber-physical system (CPS), 87, 89, 93, 102–106, 108–112
Cybersecurity, 112

D

Data analysis, 2
Data analytics, 184
Data collection, 191
Data discovery, 185, 187
Data generation, 193, 194
Data genetics, 192
Data governance, 200
Data management, 184
Data mining, 185, 188
Data privacy, 199
Data quality, 197, 198
Data role in health care system, 184
Data transmission, 257

263

Index

Data visualization, 194, 195
Data-driven smart ecosystem, 70, 71
Database, 17–19, 21, 162, 164–165, 171
DDS 47, 50
Decentralization, 84, 94, 97, 98, 166
Decentralized peer-to-peer network, 148–149, 153
Deep learning, 89
Descriptive analytics, 184, 185
Devices, 26–30
Diagnostics analytics, 184, 185, 186
Digital signature, 95, 150
Distributed denial-of-service (DDoS), 93, 162
Distributed ledger, 149, 154
Dock, 98
Driverless automobiles, 2
Driving efficiency, 8, 13
Drone Food Request Table, 244
Drone optimization, 241
Drones, 240
Dynamic traffic light sequencing, 21

E

e-call vehicle service, 17
e-governance, 87, 90, 91, 141
e-healthcare, 69, 92
e-prescribing, 92
Ease of doing business (EODB), 89
Easily applicable, 69
Education 83, 84, 85, 92
Education and encouragement, 70
Efficiency, 70
Elections, 162–163, 167, 169
Electronic control unit (ECU), 125
Electronic health records (EHR), 186
Electronic map (EM), 120, 121
Electronic retrieval and data mapping, 192
Electronic toll collection (ETC), 20–21
Emergency warning system, 85
Empowerment of patient, 70
Encryption, 84, 94, 97, 98
Energy, 32, 34, 35, 37
Energy-consumption, 83
Enhanced quality improvement, 70
Environment, 26, 28–30, 32, 33
Environment perception, 122
Equality, 70
ESRT, 254
Ethereum, 85, 100
Ethics, 70
Eucalyptus
Examining the patient, 186

F

Face recognition, 89
FDM, 44

Financial, 197
Financial services 84, 90
FinTech, 90, 99
Fog, 46
Forward propagation, 130
Framework, 26–28, 31–34
Frequency spectrum, 12

G

General data protection rule, 93
Global positioning system (GPS), 118
Global traffic management system, 118
Governance, 162–163, 165, 167, 169, 171
Government and private cloud, 76
Government-to-business (G2B), 91
Government-to-citizenry (G2C), 91
Government-to-employee (G2E), 91
Government-to-government collaboration (G2G), 91
GPS, 11, 17
Green smart teaching-learning environment, 65
GSM, 45

H

Hadoop Distributed File System, 51
Health care, 184–202
Health care information system, 184–202
Health management, 197
Healthcare, 83, 84, 85, 87, 92, 93, 98
Heat map, 91
High-definition map (HD map), 120, 121
High-occupancy vehicle (HOV), 23
High-speed, 26, 30, 31, 37
HIPAA, 199
Home lighting controller, 211
HTTP, 47
HTTPS, 50
Hybrid delivery process, 245

I

IaaS, 47
ICT, 221
Immutability, 152
Impact of COVID-19 on education, 66
Indian healthcare system, 196
Industrial automation, 104, 111
Industrial internet of things, 139
Information, 26, 27, 31, 34
Information collection, 2
Information communication technology (ICT), 87–90, 91, 94, 96, 97, 99
Infrastructure, 1, 2, 5, 7, 11, 16–17
Infrastructure as a Service see IaaS
Intelligent, 83, 89

Index

Intelligent driver model (IDM), 9–10
Intelligent system, 209, 210
Interchange, 98
Internet, 26–28
Internet of everything, 45
Internet of medical things, 139
Internet of things (IoT), 2, 4, 6, 8, 10, 12, 14, 16, 18, 20, 22, 24–31, 41, 42, 44–46, 52, 53, 55, 60, 76, 84–88, 90, 91, 92–96, 98–100, 102, 105, 204, 205, 207, 209, 210, 216
Introduction, 58
Involvement of AI in various domain of SDG for smart city, 62
IoT fuzzy logic, 230, 231, 232, 233
IoT interface, 226–227
IoT sustainable smart education, 66
IR, 43
ISM, 45

J

Joint test action group, 142
Joy-driven smart city, 72

L

Lane detection, 13
Lane recognition systems, 13
Laser perception, 123
Laser route, 122
Lateral motion, 124
LDR, 207, 214
LED, 45, 205, 207, 208, 209, 215
Ledger system, 84, 85, 94–96, 99; see also Distributed ledger
LiFi 45
Light subordinate resistor, 214
Local traffic management system, 127, 128
Longitudinal motion, 124
LoRa, 45
LPWAN, 30, 37–38
LR-WPAN 45, 47

M

Machine learning (ML), 30, 88, 89, 91, 92
Machine learning (ML) for sustainable smart education, 66
Machine to machine, 26, 28, 34–35
Machine vision, 13
Map coordinating, 121
MATLAB, 247
MBSP, 91
Media access control, 44
Merkle root, 154
Microcontroller, 207, 208, 209, 211, 215, 216
Middleware, 137, 154

MIMO, 15–16, 24
MMDR, 254
Mobile computing, 26
MQTT 47, 49
MRTP, 254
Multimedia data transmission, 257
Multi-sensor combination, 14
Multi-station shared vehicle system, 18–19

N

National information center, 91
Network, 26–28, 30
Network layer, 106–109
Network security, 102
Neural networks, 13
NodeJS, 168, 170–171
Nonce, 154
Non-invasive attacks, 142, 143

O

Object recognition, 122
Ontology, 246
Ontology editors, 246
Optical route, 122

P

PAN, 45
Path planning, 124
Patient record, 199,
Peer-to-peer (p2p), 84, 85, 94, 95, 97
Pegasus attack, 93
Performance metrics, 255
Phishing attack, 93
Physical devices, 102, 106, 107
Physical layer 106, 107, 109, 110,
PIR sensor, 203, 204, 206, 207, 208, 211–215
Platform as a service (PaaS), 147
Positive impact of COVID-19 on education, 68
Predictive analytics, 184, 188, 189
Preferred Reporting Items for Systematic Reviews and Meta-Analyses (PRISMA), 86
Prescriptive analytics, 184, 190, 191
Privacy, 27, 29, 37, 199
Process of machine learning, 67
Proportional integral derivative (PID), 126
Protected health information (PHI), 199
Protocols, 28, 29, 31
Public reporting, 198

Q

QoS 47, 48
QR Code, 43
Quantum technology, 89, 99

R

Radar route, 122
Radio frequency identification (RFI), 17
Radio transponder, 20
Real-time applications 102
Real-time multimedia transmission, 256
Redis, 127
Reliability, 163, 167–168, 171, 255
Reliability mechanism, 255
Rest API 47
Reversed logistics, 230
RFID, 17, 21, 24, 43, 53, 103, 204, 205, 215
RFID distribution, 229
RFID in SCM, 225
RFID integration, 227, 228
RFID operations, 228, 229
RFID purchasing, 229
Ripple 85, 100
Robot Food Request Table, 244
Robots, 240
Routing, 124
RTP, 253

S

SaaS, 47
SAP, 88
SARTHI, 91
SC challenges, 219, 220
SC improvements, 226–227
SC managers, 219–220, 228
SC processes, 219, 223, 226–227
Scalability, 167–171
Security, 27, 28, 29, 31, 32, 35, 165, 170–171, 199
Security number, 200
Security threats, 109
Self-driving car(s), 2–5, 16, 23–24, 117; *see also* Autonomous driving
Semantic analysis, 89
Semi-invasive attacks, 142, 144
Sensor(s), 26, 28, 29, 30, 33, 34, 102, 104, 106, 108, 110, 112, 205, 207, 208, 209, 210, 211, 212, 214, 215
Sensor fusion, 12
Sensor processing unit (SPU), 129
Sensor transmission control protocol, 254
Sidechain, 167, 169
Simulink, 247
Simultaneous Localization and Mapping (SLAM), 12, 122
Smart agents, 19
Smart car, 8–9, 11–12, 14–15
Smart city/smart cities, 1, 3, 5, 7, 9, 11, 13, 15, 17, 19, 21, 23, 25, 83–100
Smart city and AI increasing from 2014 to 2021, 61
Smart cities mission, 88
Smart ecosystem, 70
Smart finance, 73
Smart fitness, 68
Smart governance, 75
Smart home, 206, 216, 217
Smart live-style, 72
Smart mobility, 16, 83, 92
Smart room, 207, 210, 216
Smart sensor networks, 17
Smart sensors, 16–17
Smart vehicle technology, 8
Smart-device, 83
Smartphones 6, 52, 59, 69, 83
Social engineering, 93
Software as a Service *see* SaaS
Software defined networks, 45
Software-defined networking, 111
Spatial modulation (SM), 15–16
SQL, 52
SSH, 51
SSL, 51
STCP, 254
Storage, 164, 168, 171
Supply chain management, 218
Supply-chain, 85, 96
Sustainability a strategic plan of smart cities, 59
Sustainable development goals for smart city, 61
Sustainable smart learning, 64
Sustainable supply chain, 222, 235
Sustainable transport for smart cities, 63

T

TBSP, 91
TCP 47, 49
T-CPS, 20
Teaching-learning environment, 64
Technologies used for smart education, 65
Technology-based smart city, 60
Telecommunication, 26
Teledentistry, 92
Telemedicine, 83, 92
Third-party 85, 94
Time-stamped, 97
Traffic management centres (TMCs), 8
Traffic signals, 7–8, 13
Transactions, 162–164, 166, 170
Transformation toward urbanization region wise from 1950 to 2050, 74
Transparency, 162–163, 167–168, 170
Transport protocol, 255
Transportation system, 1–3, 5, 7–11, 13, 15, 17–19, 21, 23–25

Index

Transportation, 83–85, 87, 88, 89
Transportation-cyber physical system, 20
Types of machine learning, 67

U

UDP, 47
Ultrasound, 12
Ultraviolet, 12
Unsupervised calibration, 123
USB connector, 208
Utility of information and communication technology, 76

V

Variable message signs (VMS), 18
Vehicle communication, 1
Vehicle control, 6, 18, 20, 125
Vehicle navigation, 118–120
Vehicle routing problems, 241
Verifiable, 97
Virtual Machine, 54

Visualization, 186, 194, 195
Votes, 162–163, 165, 167–168, 170

W

WannaCry ransomware, 93
Wi-Fi, 206, 207, 215
WiFi, 6–7
WiMax, 44
Wireless, 85, 87
Wireless fidelity, 43
Wireless local area network, 44
Wireless sensor network (WSN), 27, 137
WMAN, 44
Worldwide internet user, 60
Worldwide path planning, 121; *see also* Universal Path Planning
WSN, 252, 253, 255

X

XMPP, 47

Printed in the United States
by Baker & Taylor Publisher Services